Do Cats have Belly Buttons?

Do Cats have Belly Buttons?

And Answers to 244 other Questions
on the World of Science

Edited by
Paul Heiney

Illustrations by Bill Ledger

SUTTON PUBLISHING

First published in the United Kingdom in 2007 by
Sutton Publishing, an imprint of NPI Media Group Limited
Cirencester Road · Chalford · Stroud · Gloucestershire · GL6 8PE

British Library Cataloguing in Publication Data
A catalogue record for this book is available from the British
Library.

Hardback ISBN 978-0-7509-4645-2

Typeset in GillSans Light.
Typesetting and origination by
NPI Media Group Limited.
Printed and bound in England.

Contents

Introduction

Let's leave belly buttons out of it for the moment and remember another old proverb about cats which says 'curiosity *killed* the cat'. That one has always bothered me. What could ever be risky about being curious? Surely there can't be anything more exciting than forming a question in your mind, turning it over till you've looked at it from every angle, deciding you don't know the answer and then, armed only with your curiosity, go seeking the answer? There's nothing foolhardy in being curious. Surely one of life's most rewarding journeys are the steps we take from question to answer?

But since this is a book of science questions and answers let's be scientific about this and ask ourselves if there is any evidence that having a curious mind does you harm? As a sample of the population, let us take the tens of thousands of people who, a few years ago, phoned or emailed an organisation called 'Science Line'.

Science Line was set up with a simple purpose: it was there to answer all manner of queries from deadly serious to wildly zany, on every branch of science from cosmic physics to microbiology. It fell to a gang of enthused, young scientists to try and answer them. If they couldn't find the answer they found someone who could, and in that way the most humble enquiry often found its way onto the desks of some of the top scientific brains, who were delighted to help. Suddenly science didn't just belong to the people who knew all the answers, it could be shared equally with those who posed the questions.

And was anyone harmed by asking questions of Science Line? I don't think so. I have no proper scientific proof, but I doubt if any of the 500 people who emailed every week (adding up to tens of thousands over the years) came to any harm. So we can reasonably assume, if not scientifically prove, that no amount of curiosity ever killed any questioner, even a cat.

But, of course, what the proverb is really suggesting is that if you are too interested in things which are none of your business, you could be in danger. I can see some sense in that in certain circumstances. But there is nothing to be found anywhere in science into which we have no right to stick our noses. Science is everyone's business – it describes *our* lives, *our* world, *our* universe – and Science Line made it that bit easier for everyone to share and understand it.

Although some questioners sought deep insight into the mysterious depths of quantum theory or molecular movements, others were happy simply to ask why squirrels have bushy tails, or even if cats have belly buttons? Indeed, the puzzled soul who asked whether cows can walk downstairs, kindly provided us with the title of the first book in this series.

For this second book, I have revisited the vast database of questions and answers that survived after Science Line finally closed for business, when the funding ran out. There seems to be no end to it. It sometimes feels as if I have not only struck gold, but the deeper I dig the more gold there is. If you thought *Can Cows Walk Downstairs?* covered pretty much everything you needed to know, in this book you will find areas of science we have not visited before, such as the science of bubbles or the movement of ping-pong balls, a journey to the centre of the Earth or making toast in a thunderstorm.

Once again, I must thank those who asked the questions, and those who answered them. They were: Sian Aggett (Biology), Alison Begley (Astronomy and Physics), Duncan Kopp (author of *Night Patrol*), Khadija Ibrahim (Genetics), Kat Nilsson (Biology), Jamie McNish (Chemistry), Alice Taylor-Gee (Chemistry), and Caithlin Watson – as well as the numerous distinguished experts whose knowledge they drew upon when their own was stretched to its limits.

And finally, may I reassure you that no cats were harmed in the preparation of this book. And to discover the truth about their belly buttons, read on.

Paul Heiney
2007

1

THE
Human Body

Big Ears to Big Sneezes

Funny Bones to Laughing Gas

Belly Buttons to Hairy Eskimos

Happy and Sad

Sweet Dreams

Why the Difference?

Just Wondering . . .

BIG EARS TO BIG SNEEZES

Do **people** with **sticky-out ears** have better **balance?**

It's true that our ears allow us to keep our balance, but I think you have got hold of the wrong end of the stick about how they actually achieve that. It is nothing to do with the size of the external ear, the pinna.

A special part of the inner ear, called the vestibular apparatus, helps the body to cope with changes in position. This structure contains hair-like cells which wave about in the fluid inside the inner ear and connect to lots of tiny nerves. These all work together to tell the brain what position the body is in and whether it is moving or not. When the information from these hairs is at odds with messages going to the brain from our eyes, we can suffer from motion sickness such as car or sea-sickness.

However, these messages originate in the inner ear, not the outer ear, and so the size of your external ears make no difference to your balance at all. Unless, of course, they're so large that you trip over them.

Can **sound** hurt your **ears?**

Parents are always nagging their children to 'turn it down!' and not just because it is annoying. Sound can most certainly damage your ears. Hearing tests on gunners and people who have worked near jet engines show they can no longer hear high-frequency sounds, and have difficulty hearing normal speech. Even wearing headphones with the sound turned right up for long periods of time can cause some damage to hearing.

Sound travels in waves through the air but, unlike waves on water which we can see coming towards us and duck away from if we have to, there is no simple way to see a damaging sound wave coming at you. The only way to measure the power of a sound wave is to use a microphone to convert the sound waves into electrical waves, and measure the voltage produced by the microphone. The usual scale measures the loudness (sound pressure level) in decibels (dB), on a special sort of scale where 40 decibels is ten times louder than 20 decibels and 60 decibels is a hundred times louder than 20 decibels.

0dB, which is reckoned as being the threshold of hearing, represents the quietest sound we could ever hear – the sound of an empty building on a quiet night in the country. You would probably play music at about 40dB to keep you company while you do your homework. Traffic noise at rush hour in a busy city might reach 80dB, and the threshold of pain would be 120dB – roughly what you'd hear if you stood at the end of the runway when a jet aircraft took off.

Hearing loss occurs when there is severe damage to the structure of the highly sensitive inner ear, particularly the hair cells which transmit vibration to the brain for recognition. The first effects will be the loss of high frequencies which are important as they enable us to recognise the difference between similar words, such as thrill and sill. In severe cases, conversation begins to sound like a continuous mumble.

Remember, damage starts at about 80db – the roar of busy traffic. Rock concerts can hit 115, a passing ambulance 125 and a shotgun fired close to you 165.

What is the **lowest intensity** of **light** the **human eye** can **detect?**

All it takes is one single photon. A photon is complex to define, but you can think of it as being a particle of electromagnetic energy. Light consists of streams of photons, the smallest particles of light thought to exist.

Light is detected by cells in the retina at the back of the eye, called rods and cones. Rods are more sensitive than cones, and a single photon of light is enough to cause a rod in the human eye to fire up and send a message to the brain 'photon received!' How bright is a photon? Roughly equivalent to a single candle viewed from one mile away – not much.

Why do we **need** two **eyes?**

A pair of eyes produces binocular vision, which means that although our brains receive a different image from each eye, we only 'see' one image. Both in humans and animals, having two eyes is useful because it provides a larger field of view, and reduces the risk of becoming disabled following damage to one eye. It also allows stereoscopic vision so we can see things in three dimensions.

The placing of the eyes is important: in the animal world predators often have their two eyes placed on the front of their head to maximise this overlap of retinal images, giving excellent stereoscopic vision, allowing them to judge distances accurately and locate their prey. On the other hand, their prey tend to locate their eyes on the side of their head which reduces stereoscopic vision but gives a much improved all-round sight to help detect nearby predators.

Why do you get **dizzy** standing **on top** of a **tall building?**

Because your eyes are used to seeing the ground somewhere near your feet. If it suddenly spots them somewhere else, it starts to get confused. This mental confusion causes the sensation of dizziness. Because the perspectives are wrong, the confused brain starts to over-correct and, apart from the feeling of dizziness, it can also lead to a great feeling of anxiety.

If you **swapped** your **eyes** over so that your **left eye** was in your **right socket** and **vice versa**, would you see the world with **two halves** which didn't **match up?**

It's quite likely the brain could sort this out without you even noticing. After all, we're already seeing things upside down and back to front.

Images from our eyes are transmitted via the optic nerves to the brain through fibres which are divided into two bundles. One bundle contains fibres originating from cells on the temporal side of the eye – the same side as the ear – the other bundle contains fibres originating at the nasal side of the eye nearest the nose.

7

From here onwards you might want to draw a diagram of a head with two eyes, and a brain with two hemispheres – it helped me.

The fibres originating from the temporal side go back to the hemisphere of the brain on the same side of the head as the eye in which the fibres originated. The nasal fibres cross over and go to the opposite hemisphere.

Simple lenses produce images that are upside down, and the eye does the same. This means that if you imagine a human figure viewed by an eye, the image is inverted so that the head is at the bottom and the feet at the top. Also, the left side would be on the right, and the right side on the left. The inversion is actually a rotation of 180 degrees. This means that the image of the left side of the scene is formed towards the right side of the retina. So, in the case of the right eye, images formed from objects on the right side of the direction of gaze land on the left side of the retina – i.e. nasally. These nasal images would lead to neural signals that are transmitted to the left hemisphere. Points imaged on the left of the direction of gaze would be imaged in the right side of the retina, and signals produced would be transmitted to the right hemisphere.

So, images on the left side of either retina produce signals that are transmitted to the left brain hemisphere, and images on the right side of either retina are sent to the right hemisphere. The crossing of these signals to the two hemispheres is what allows us binocular depth perception.

About 70 per cent of the total number of fibres originating in one eye cross over, while 30 per cent remain uncrossed and go to the same side. So, if you put your right eye in your left socket, but attached it to the optic fibre originally in the left socket, and the same on the other side, then the brain would still receive images exactly as before and therefore nothing need be fixed.

However, if you put your right eye, plus the connecting optic fibre, in your left socket, then what you would find is that your peripheral vision would be in the middle of your vision, and your central line of vision on the outside. This would obviously take some fixing, but the brain can adjust to many things and it would be more than likely that your brain would eventually adapt to this without you even realising.

Don't try this at home.

How long do we **spend** in a **lifetime** with our **eyes closed** just by **blinking?**

A blink lasts about 0.3 to 0.4 seconds. We blink about 5 times a minute, every minute, for about 18 hours a day. This adds up to half an hour a day and about 5 years in a lifetime.

What are **eyebrows** for and **why don't** they **grow?**

Eyebrows are for protecting your eyes by deflecting water running down the forehead, and they're used in conveying facial expressions. They don't grow longer because the hair follicles are genetically programmed to stop after about a centimetre of growth which, I guess, is the reason you don't need your eyebrows trimmed every time you get your hair cut.

If your **eyebrow** is **shaved**, how long does it **take to grow?**

Human hair grows about 23cm a year and eyebrows tend to be about 1cm long. So, if you shaved your eyebrows, it would take about 17 days for them to grow back.

What are the **black floaters** you **sometimes** get in your **eyes?**

What you are seeing are small particles of debris swilling around in the vitreous fluid which fills your eye. Don't worry

– in the vast majority of cases they are entirely harmless and most people will experience them to some degree if they stare hard enough and long enough at a blue sky or a blank wall.

The floaters might consist of very small quantities of blood or tissue which has become detached from the retina. But more usually they are part of the ageing process of the vitreous fluid, which is why floaters tend to be a problem for older people. They can also appear as streaks, clouds or spiders' webs. Incidentally, it is not the floater itself which you are seeing, it is its shadow as it passes across the retina.

Why do people **sneeze?**

We sneeze to clear irritating material from our upper air passages. This can be anything from dust, pollen or snuff, to excess mucus blocking the nose when we have a cold or hay fever. Pain receptors in the cells lining the upper respiratory tract are triggered by the dust or mucus and instruct the medulla (the base of your brain) to make you sneeze.

The sneeze itself is just a very powerful out-breath which propels the air at up to 100mph and can discharge 5,000 water droplets, all loaded with bacteria. A large number of muscles suddenly come into play just before a sneeze, including stomach muscles, the diaphragm, the muscles in your throat and, of course, your eyelids because it is almost impossible to sneeze with your eyes open.

A sneeze starts with a closing of the vocal cords until the pressure in the chest has risen, and then the air is suddenly allowed to escape upwards into the back of the nose by a soft palate. But the 100mph which the sneeze reaches is nothing compared to the 600mph at which a cough can travel.

When you **sneeze**, why do you see **bright lights?**

Remember how close your eyes are to your nose? As that blast of air is expelled at 100mph (see above) the eyeball is pressed hard against your eyelid, which always closes during a sneeze.

The eyeball is filled with a jelly-like substance, so any pressure on the front of the eyeball is transferred to the back where it falls on the retina. The retina cells are not only sensitive to light (they are what enable you to see) but they are also sensitive to pressure and will stimulate the nerves just as if light was falling on them. So, when you sneeze, the pressure on the retina presses on the nerves, which stimulates them, and the brain interprets the messages as being the result of light. This is quite reasonable since most of the time nerve messages from the eyes *are* the result of light falling on the retina.

By the way, you can also see lights which aren't there by closing your eyes and pressing gently on your eyeball. Migraine sufferers often complain of seeing bright lights when they get a migraine headache. This is because the blood vessels in the area of their head around the eye contract, so the blood is at higher pressure than normal, and presses on the eyeballs.

FUNNY BONES TO LAUGHING GAS

What's the
funny bone?

Don't laugh when I tell you, but the funny bone is not a bone at all. It is a nerve which runs through a groove in a bone called the ulna, one of two bones of the lower arm. At the elbow, this bone and its nerve are very close to the surface and stick out, making it easily hurt by knocking or bumping. If you manage to hit your elbow in precisely the right spot, it sets off the tingling or prickly sensation which we describe as having hit your funny bone, although it's very rarely a laughing matter.

What you have actually struck is your ulnar nerve which controls feeling in your fourth and fifth fingers and plays a part in wrist movement. Because the nerve itself is being hit, you get a very painful physical reaction, making it hurt an awful lot more than you might expect. When things hurt a lot you get very emotional, which means you either laugh or cry. Quite often you can laugh and cry at the same time, and so people have come to call it the funny bone because it makes you laugh (or cry) when you hit it.

Don't confuse the funny bone with the humerus (not humorous) which can be found in the upper arm. As far as we know, it has no sense of humour at all.

How many **bones** are there in the **human body?**

The skeleton of an adult contains 206 distinct bones, and these are contained in two separate systems: the axial skeleton, which you can think of as the main trunk of your body, and the appendicular skeleton which is the arms and legs. There are 26 bones in the vertebral columns, 8 in the cranium, 14 in the face, 7 other skull bones, 25 in the sternum and ribs, 64 in the upper limbs and 62 in the lower limbs.

What is the **smallest bone** in the **body?**

The smallest bone is to be found in your ear and it is called the stirrup bone, so-called because it is shaped like a horse

rider's stirrup. It is one of three little bones in the middle ear and measures from 2.6 to 3.4mm in length and weighs from 2.0 to 4.3mg, which makes it about the size of a grain of rice. Because of their shape, the other two middle ear bones are known as the hammer and anvil, and all three sit within the same cavity. When sound waves collide with the eardrum, it is these three small but vital bones which transmit the vibrations to the inner ear where sensitive nerves in the cochlea translate them into nerve impulses which are sent to the brain.

What am I **made of?**

Mostly you're a bag of water. Seventy per cent of the body is water. Since water consists of atoms of hydrogen and oxygen, you could say those are the two elements of which you are made. Of course, there are other elements such as carbon, nitrogen, phosphorous etc. Altogether there are 60 different elements found in the body. The most abundant is oxygen and a 70kg person would have 43kg of oxygen in their body, 16kg of carbon, 7kg of hydrogen, 1.8kg of nitrogen and 1.0kg of calcium. Those are the top five.

By the time you get to the bottom of the list the elements are becoming scarce. The least found is tungsten (20 micro-grammes) and vanadium, thorium, uranium, samarium and beryllium are present in only slightly larger quantities.

However, the atoms of these elements are made up of just three particles: protons, neutrons and electrons, so to say that you were made entirely of these three would be another equally valid answer. But if you want to get really serious, bear in mind that recent experiments have shown that protons and neutrons appear to be made up of smaller particles called quarks.

It's a bit like the old poem:

> Big fleas have little fleas
> which sit on 'em and bite 'em
> but little fleas have smaller fleas
> and so *ad infinitum*.

Perhaps one day we shall discover that quarks and electrons are made up of something smaller, and then there will be a new answer to the question. At the moment experiments suggest quarks are as small as things get. So, if you don't want to think of yourself as a bag of water, consider yourself to be a mountain of quarks.

Who was the **first person** in the **world?**

People, as we know them, evolved over thousands and thousands of years. Scientists think that people evolved from an ape-like animal. About 2 million years ago they started to walk upright, then their brains got bigger and eventually they became like us. But this all happened very gradually and there is no specific point at which a person appeared in the world. The general name which we give to the first group of people-type animals which looked a bit like us and walked on two legs not four, is *Homo habilis*.

Why does **nitrous oxide** make you **laugh?**

Nitrous oxide, famously used by dentists until better anaesthetics were introduced, does have a reputation for leaving some people feeling giggly. In fact, in the nineteenth century, it was used as a fairground attraction where people would pay money to inhale the gas in order to behave in a silly and uninhibited way. So intoxicating was it that the poet Robert Southey wrote: 'I am sure the air in heaven must be this wonder working gas of delight.'

It was first discovered in 1793 by Robert Priestley, the great discoverer of gases, who first isolated oxygen, carbon dioxide and ammonia. But it was another fifty years after Priestley's discovery before it was used in surgery to dull pain and the senses of sight, hearing and touch.

The precise mechanism which leads to laughter is unknown but a wide range of emotions can result from inhaling nitrous oxide, among them hysteria. It is dangerous to inhale nitrous oxide unless it is mixed in the correct proportions with oxygen.

Laughing gas is used in rocket fuel and in racing cars because it supports combustion better than air, so more fuel can be burned in less time which results in more thrust.

BELLY BUTTONS TO HAIRY ESKIMOS

Why do **men** have more **fluff** in their belly **buttons** than **women?**

Men have more belly button fluff than women because they tend to be hairier. The patterns of hair on the male body tend to channel any debris towards the belly button. This debris is generally made up of dead skin, as well as fibres from the inside of your clothes. But it might also include harmful bacteria which need to be shunted off to a safe place, and the belly button provides a useful dumping ground. The extra

18

hairiness of men also means that more fibres are rubbed off the inside of your clothes, so that even more trash finds its way to that dustbin of a belly button.

Why do **men** have **chest** and **belly hair?**

It is all to do with sex. Body hair develops in sexually mature males and acts as a signal that they are ready to mate. Females have evolved to consider body hair to be sexually attractive in a potential partner. Body hair is called a 'secondary sexual characteristic' and is found throughout the animal world as well as in humans. It might be the bright plumage of a bird, or the mane of a lion. In humans it might be the development of facial and body hair in males, and the development of breasts in females.

So body hair sends the signal that a man is sexually mature and able to reproduce. Also, it makes it absolutely clear which are the males and which the females to avoid wasting time and energy on fruitless courting rituals.

The pattern in which this hair develops is genetic, with certain racial types being excessively hairy and some, such as the American Indians, never developing any body or facial hair at all. However, generally speaking, chest hair grows in a line which runs from each armpit to the groin.

19

Are Eskimos **hairier** than other **people** to keep them **warm?**

It's a nice idea, but there doesn't seem to be much evidence that Eskimos are hairier. Probably the thick fur clothes they wear means that thick hair is unnecessary. The combination of their clothing, lifestyle and behaviour keeps the air temperature next to their skin comparable to that of people in warmer climates.

The exception to this is the hands and feet where even the thick Eskimo mittens and boots are not enough to keep them warm. Consequently, Eskimos have better blood circulation than other races to spread heat to their entire body. Also, Eskimos have a short, compact body shape with quite short arms and legs, which also helps to minimise heat loss. People from hot countries such as Africa tend to be much taller and thinner with longer arms and legs.

When you **pull out** a **grey hair**, why does another grey hair **replace** it?

Hair consists of long filaments of protein, called keratin, which are secreted by the hair follicles of the skin. The colour of your hair is determined by the concentration and depth of the pigment, melanin, in the hair shaft. Once the cells which produce the melanin, the melanocytes, become inactive or die, your hair goes grey or white. Once the melanocytes have died, they will never produce a coloured hair again. So no matter who many times you pull out that grey hair it will always grow back grey. But look on the bright side, pulling out a grey hair will not result in several more grey hairs growing in its place, only one.

Does **hair** grow after you **die?**

This may be one of those urban legends – rumours which become accepted as fact. There is also a rumour that our fingernails continue to grow after we die. Neither of these is true. Follicles need a blood supply in order to produce hair and as soon as that stops so does the growth. Same with nails.

21

The truth is that after death a corpse starts to lose water and dehydrate, and as it does so the skin shrinks and pulls back. At first glance, this could give the impression that the hair and nails are growing, but the truth is that the skin is receding.

HAPPY AND SAD

Why do we **laugh** and what causes **laughter?**

The exact mechanism that causes us to break into laughter is not known. In fact, it is one of the least understood of all the brain's workings and no specific part of the brain has been found where we could say with certainty that this is the source of laughter. We know, though, that it can lead to a major upheaval of the body's system, including changes in facial muscles, heaving of the abdomen, movement of the limbs, elevation of the shoulders, as well as a hearty vocal chuckle, or even guffaw, which we recognise as a laugh. There is also the emotional elevation which can transform attitudes within seconds. Remember, though, that there is also the nervous laugh which comes in useful when we are uncertain of ourselves.

Some psychologists say that laughing is just another part of our vocabulary, a way of expressing enjoyment without having to string together descriptive words. It is true that when you laugh at someone's joke, it is a far more efficient mechanism for expressing delight than a hundred words could convey.

Laughter is something that evolved in man. If you tickle apes they are not capable of a recognisable laugh, but they do pant and anthropologists believe that the panting turned to laughter at some stage in our evolution. You laugh more between the ages of five and six than at any other time in your life. It's not bad exercise either: estimates say that laughing 100 times is as good as 10 minutes on a rowing machine.

Why do we **cry?**

If you're a baby, you cry to communicate: to let your parents know you're upset, hungry or hurt. However, don't expect to see a baby cry tears for the first few weeks of its life since the tear-producing lacrimal system takes some time to develop. Instead, expect a dry cry. Adults, of course, cry less often than babies because they have other ways of getting their message across. On average, women cry 67 times a year, men only 17.

Crying is a major physical exercise for not only are you releasing tear drops – protein-rich, salty water – but you may make the involuntary gasping sounds that often accompany tears, as well as muscle spasms that may tighten the chest and cause the lungs to inhale and exhale deeply.

There are three types of crying. The first is happening all the time on a small and unemotional scale. It is essentially an

eye-washing operation where the tear fluid is spread across the eye by blinking and kills off germs and removes dirt, preventing the eyeball from drying out. This is called basal crying.

Reflex crying happens when you get dust in your eyes, or some other irritant. Your eyes fill with tears in an attempt to wash the intruder away.

Much more difficult to explain is emotional crying, which is not restricted to sadness of course – you can just as easily cry if you win a jackpot. One theory says that when we cry due to distress, it is a signal to others that we are in need of help, or to display our true feelings. Incidentally, emotional tears are different to other tears – they contain 25 per cent more protein and a mixture of hormones.

The truth is that no one knows precisely why we cry. The ancient philosophers thought it was a mechanism for cleansing the emotions and for sending signals we have no other way of expressing.

What **effect** does **colour** have on your **mood?**

Colour affects people in different ways. In some recent experiments, people tended to feel more tired after filling in a form printed on red paper than they did after filling one in on green paper. Workers in a red office apparently made fewer mistakes than workers in a green office, but they felt much more stressed out. Young children seem to be happier in a pink room, rather than a blue one. Some weightlifters say they can perform better in a blue gym.

It is impossible to say why, but there does appear to be a general consensus on the significance we attach to colours. Black, for example, symbolises authority and power over others – which may be why the black suit is a powerful way to dress for office confrontations. (Remember, though, the Devil is usually robed in black.) White implies innocence, and is the bridal choice of colour, while green is calming and may be the reason why actors retire to the 'green room' before going on stage. Brown is sad, purple suggests wealth.

Can **stress** make you **sick?**

Scientists are beginning to realise that there is a strong link between our emotions and our immune system. After the stress of exams or an upsetting incident like losing a friend or relative, it is common for people to come down with colds, or sometimes a more serious illness. On the other hand, people who are happy and in supportive, loving environments often recover quicker from illness – happy people get better sooner. So our mental well-being does seem to have some influence on our health.

The brain links up with the immune system in various ways. Stress activates the hypothalamus, which is the part of the brain controlling hormone release from the adrenals and the pituitary gland, and these hormones help us to cope with emergencies. The autonomic nervous system is also revved-up by stress. This system controls the major organs and muscles of the body without us being aware that it is happening – the breathing in and out which you are doing now is triggered, in part, by the autonomic nervous system, which also gives us that classic 'fight-or-flight' response to cope with emergency situations. It's not something we have any control over – we can't stop ourselves being afraid. The brain also mobilises the organs of the immune system – the lymph nodes, the spleen, thymus and bone marrow. As a result, the number of B cells which produce antibodies, the T cells which fight on our behalf, and the macrophages which remove foreign debris, can be regulated. So, all manner of internal events take place as a result of stress and in humans these may be triggered by everyday pressures of work and travel, or lack of sleep and depression. If such demands continue, people become chronically stressed and the body's resources are directed towards its current emergency rather than keeping its long-term immune system fully working. That is how stress lowers our immunity and allows us to get sick.

What causes the feeling of 'butterflies in the stomach' when you are nervous or excited?

The butterflies themselves are easy to explain; they are small muscle contractions in the digestive tract. The gut's muscles are normally well coordinated but when disrupted by stress they contract in an odd way that we start to notice, and these are what we call 'butterflies'. *Why* they happen is a bit more difficult to understand.

Our entire digestive system, from mouth to anus, has its own nervous system – the enteric nervous system – which lines walls of the digestive system and is connected directly to the brain. Now, imagine something really dangerous is happening – you are being chased by a tiger. At this particular moment, digestion is not a priority so there's no point wasting blood and oxygen on it, especially at a time when heart, lungs and muscles need all the boost they can get. So the brain decides to shut down the gut and this seems to wreak havoc on normal gut functions. This can cause many symptoms. Some people might develop diarrhoea, vomiting and stomach cramps. Some will experience a strange sensation as the stomach muscles have a brief panic at being told to shut down, and that is the feeling of butterflies in the tummy.

Why do we remember the **scary** parts of **movies** better than the **happier** bits?

Researchers in California have suggested that a structure in the centre of the brain called the amygdala may be the storehouse of these terrifying images. This is the part of the brain which decodes emotions and detects threats.

They did an experiment in which researchers showed a group a set of disturbing images (pictures of torture), and then another set of emotionally neutral images. At the same time they monitored which areas of the volunteers' brains became active when they looked at these pictures. They did this using PET, Positron Emission Tomography, which allows them to follow the flow of radioactive material injected into the blood as it passes through the brain.

When questioned three weeks later, the volunteers always remembered the violent images most clearly. The research showed that the recall was most vivid when the brain activity in the amygdala was at its greatest.

It is also thought that the amygdala might be an important storehouse for emotional memories. Damage to the amygdala in one patient meant they could no longer recognise the look of fear on a human face. Another patient, also with a damaged amygdala, was no better at recalling frightening events than those that were not upsetting. So it seems the amygdala is where the more vivid and frightening memories are stored.

use of its position within the brain, the amygdala is positioned to capture much of the sensory information that the body receives and so it is an ideal mediator between the outside world and how we react to it. The amygdala may well keep the body on the lookout for potentially dangerous situations so we can escape from them quickly – scary bits of movies included.

SWEET DREAMS

Why do some **people** need less **sleep** than others?

Millions of years of evolution has given us a built-in biological 24 hour clock linked to the daily rotation of the Earth. This internal clock is in charge of all our metabolic processes and decides when it is time for us to feel sleepy and when we are ready to wake up. Your own internal clock is deciding for you how much sleep your body needs.

But no two people are alike in their sleep needs. Each of us has a specific daily sleep requirement and if this is not met, we create what can be thought of as a sleep debt. Sooner or later that debt must be repaid. You can't cancel it, and it won't go away. If you're short of sleep, sooner or later you are going to have to catch up. The powerful brain mechanism that regulates the daily amount of sleep is called the sleep homeostat which ensures that most people will get the amount of sleep they need, or close to it, by sending familiar signals such as drowsiness, or drooping eyelids. You can fight these signals for a certain amount of time, but sooner or later the sleep homeostat is going to win.

As far as is currently known, nothing can change an individual's fundamental daily sleep requirement, but you may be programmed to need more or less than someone else. So some people will sleep longer than others. There are no rules.

What causes us to feel **sleepy?**

As we've just read, the size of your sleep debt determines whether or not you fall asleep. If your sleep debt is zero, sleep is impossible. If your sleep debt is very low, only a small amount of stimulation is required to keep you awake. If your sleep debt is very large, no amount of stimulation can keep you awake.

Think of your sleep debt as a very heavy load which you are carrying with the help of two companions. Together, the three of you can hold it up. One of your companions is pretty strong – this companion is your biological clock. The other companion is not quite so strong, and represents external stimulation such as noise, light, excitement, anger, pain, and so on. If one of your companions drops out, you and the other may be able to manage. If both companions drop out and you are left alone, you absolutely cannot hold up the heavy sleep debt and you are crushed. In other words, you cannot stay awake no matter how hard you try. It is usually easy to stay awake and alert if your stronger companion, the biological clock, is helping you.

If you believe that boredom, a warm room, or a heavy meal causes sleep, you are completely wrong. If boredom, a warm room, or anything else seems to cause you to feel drowsy, you have a sleep debt and you need to be stimulated in order to stay awake. If you frequently feel sleepy or drowsy in any dull or sedentary situation, you almost certainly have a very large sleep debt.

What happens, **chemically** speaking, when we fall **asleep?**

Over an average lifetime, we can spend as much as 24 years in bed, yet science has only a poor understanding of what sends us there and what causes our eyes to close and our bodies to pass into sleep.

We know that the pineal gland, located at the base of the brain, seems important because here a chemical called melatonin is produced which enters the blood stream to regulate the cyclical sleep pattern. In the early twentieth century, researchers began to think that sleep was caused by natural chemicals – sleep substances – which accumulated in the brain, but recent research suggests that these chemicals only modify and regulate sleep, rather than cause it.

Sleep happens at five distinct levels, the fourth being the deepest. The fifth stage, at which rapid eye movement occurs, is called REM sleep and is much lighter. There is no snoring during REM sleep so snoring is a sign of being deeply asleep. Our bodies produce important chemicals during sleep; in both adults and children the output of human growth hormone surges during the first three hours of sleep. Nobody knows why this occurs, and it does not mean that people grow while they are asleep. A neurotransmitter called serotonin features during dreaming, and this occurs only in REM sleep. However serotonin is constantly produced in our bodies and not just when we are asleep. So there's no real answer to the question. Sleep remains hugely mysterious, but always welcome.

What is **'sleep'**, as in the weird yellow **gunge** in your **eyes** in the morning?

There is no proper medical description for 'sleep', although everyone knows what it is. But we know what it consists of. It is crystallised deposits of lysozyme, saline and protein which collects in the corner of your eyes, and the first daily task is generally to remove it with a rub of the eyes.

Lysozyme is an important enzyme which destroys bacteria cell walls and is found in the mucous membranes of the nose

and tears ducts. It is part of an important protective mechanism in your eyes, which would otherwise be a breeding ground for bacteria. In fact, eyes are an almost perfect place for bacteria to grow – they're warm, moist and, when your eyes are shut, it's dark as well.

The surface of your eye is continually washed with a solution of saline (salt) and lysozyme to remove any dirt or micro-organisms, and this collects in the corners of your eyes creating the 'sleep'. This process is occurring all the time, but when you're awake these deposits are regularly removed. When you're asleep they collect in the corners of your eyes.

It is recommended that you **don't sleep** in a room with **plants** in it. Does this mean it is **unhealthy** to sleep in a **forest?**

During the daylight hours, plants collect sunlight and turn it into chemical energy in a process called photosynthesis. One of the by-products is oxygen. At night, the process is reversed, and the plants take in oxygen and release carbon dioxide. This explains why flowers and plants are removed from hospital wards during the night as it is thought people who are ill don't need to compete with plants for oxygen although it's doubtful whether anyone has proved the flowers would do any harm. But any bugs lurking in the flowers might, which may be a better reason for removing them.

If you are thinking of sleeping in a forest and are worried about a shortage of oxygen, don't be. There's so much oxygen floating around in the air that there's more than enough for plants and humans. So, sleep well.

Do blind people **dream?**

Blind people do indeed dream and the images they see depends on how much they may have seen already. Most blind people are not completely blind – they can see some light and colour, and this will be reflected in their dreams. Someone who has been totally blind from birth will have dreams which consist only of sounds, while someone who has become blind as a child will only dream of images remembered from that time.

During a dream, the visual area of the brain's cortex is stimulated, and in the majority of blind people the cortex is unimpaired so there is no physiological reason why they should not dream. In cases of blindness where there is damage to the visual cortex, other regions of the cortex can remain unaffected allowing dreams to occur.

Blind people sometimes find it difficult to describe their dreams, and how they interpret their dreams depends on what they can remember from the time when they had sight.

What do newborn **babies dream** of?

We all dream every night, whatever age we are, and most dreams are based in familiar surroundings and about events, emotions and thoughts that have been experienced in the recent past. *Why* we dream, though, is not clear. We know that the brain needs constant stimulation to work normally and one idea is that dreaming is a way of helping to maintain brain function during sleep, to keep it ticking over. Some people think that dreams have some sort of meaning, but most believe they are just random thoughts in the brain. Until we find out what dreams are for, we won't find out why we dream.

Babies certainly dream. In fact in babies and infants REM (rapid eye movement) sleep is more than twice as long as in adults, suggesting infants dream much more than adults because this lighter phase of sleeping is when dreaming occurs. In terms of what do babies dream of, remember that even though babies do not have a highly developed sense of sight, or understanding of language, their dreams will have just as much validity for them as our dreams do for us. So babies' dreams will most likely be of bright lights and colours, fuzzy images, noises and smells. In fact, any of the physical stimulus that they experience when awake.

WHY THE DIFFERENCE?

Are **women** built to **talk** more than **men?**

If you want to joke about this, there are plenty of opportunities. For a start, here are a couple of well-known sayings:

'When both husband and wife wear trousers it is not difficult to tell them apart – he is the one who is listening.'

'The tongue is the sword of a woman and she never lets it become rusty.'

But scientists are taking it seriously. Using PET scanning techniques (see above) they set out to discover if there were any differences between the sexes when it came to both reading and speaking. And they *did* find differences.

When given a task that involved speaking, the left hemisphere of the brain (the one stronger in language skills) was activated in both men and women. But in women, areas in the right hemisphere were activated too. In other words, when it came to talking, women were using more brain power. They came to the conclusion that this extra brain power displayed itself as added emotion but not necessarily more words. In analysis of male/female conversations, though, it was rarely found that women used more words than men.

But it's possible that women do more talking, or at least mouth movements, before they're born. A doctor watched 56 foetuses in the womb and discovered that, overall, baby girls move their mouths more frequently and for longer than the boys. Boys on the other hand were more agile, moving their arms and legs more often. Baby girls do indeed learn to speak more quickly than boys, which perhaps explains their womb-bound rehearsals.

Why do men's
voices break
but women's don't?

Men have deeper voices than women and children for two main reasons. The first is that the pitch of the voice depends on the frequency of vibration of chords, and this in turn depends on the tension and length of the chords. Women and children have shorter vocal chords than males and therefore have higher pitched voices. The other reason is that a man's voice breaks during puberty due to the release of the hormone testosterone which causes an increase in the size and thickness of the vocal chords. As women don't produce as much testosterone, their voices don't break, and aren't as deep.

Why evolution has taken this course is impossible to say, but it could be argued that the deeper male pitch would be beneficial in terms of dominance. Low-pitched sounds travel further than higher noises; individuals with deeper voices are able to be heard over a larger area, and more powerfully. Males with deeper voices appear more powerful than those with higher-pitched voices, and may therefore be more successful.

Is there a reason why some people **like heat** and others don't?

It is because individuals have different 'set' temperatures, so someone can have a set point at 36°C and therefore be more sensitive to high temperatures than someone programmed at 38°C. A person with a 'thermostat' set low, switches on their regulatory mechanisms, such as sweating, at a lower temperature – so as the temperature rises they will become uncomfortable sooner than some others.

Genetic factors also play a part in whether you are sensitive to heat or not, also whether you are a man or a woman. Even though the two sexes do not have significantly different body temperatures, women tend to feel the cold more than men partly because they have a slower metabolic rate and so produce less heat. They also have less heat-generating muscle mass which is used during shivering to maintain body temperature in a cold environment. Research has also shown that women cool faster then men. Women generally have a lower blood pressue than men which means that there is less force driving the blood around the body, which explains why women are more prone than men to cold hands and feet since there is less blood reaching the extremities.

Why do people's **temperatures rise** when they are **ill?**

When you are ill and your temperature rises, you are said to have a fever. Fevers are the body's reaction to some kind of invasion. When you are ill, your body starts to fight back by releasing proteins called pyrogens which influence the temperature-control centre in the brain, called the hypothalamus. The result is a rise in body temperature. But exactly what effect the fever has on the infection is not known. Some say that an elevated internal temperature will stop the growth of any infecting bacteria, or that a higher temperature makes all internal organs work faster, meaning that we produce more hormones, enzymes and blood cells, and our blood even circulates faster.

There's an old wives' tale which says you should 'feed a cold and starve a fever'. Bad idea. Your fevered body uses up much more energy in maintaining a higher temperature, and that energy has to be replaced. So starving yourself would not be a good idea.

Why do some **sporty** people never suffer from **cramp?**

Cramp is a very unpleasant, painful and sudden contraction of the muscles, over which you have no control. Cramp can occur if you become very cold, or more commonly after strenuous exercise, and is usually caused by either lack of oxygen, salt or water in the muscles. If it happens during exercise, it is the body's way of saying 'Enough!'

But when you exercise frequently and become fit, the number of muscle fibres in your muscles increases, as does the blood supply to those muscles. The muscles can therefore exercise for longer before becoming fatigued because the increased number of blood vessels can supply more oxygen. Cramps can occur when your muscles don't receive enough oxygen to oxidise the lactic acid which is produced by working muscles. But if you are very fit, and your muscles' oxygen supply is working at full stretch, you never get to the stage when your muscles can't get rid of the lactic acid, and you don't get cramp.

What's the **difference** between **fingernails** and **toenails?**

Fingernails and toenails are pretty much the same stuff, as you might expect, but I doubt you'd ever guess that they're similar to hair. Hair is formed from thin fibres of keratin, which is a colourless protein, and nails consist mostly of hard keratin.

Fingernails are especially important in humans and primates because of the great use that we make of our hands. Nails allow us to grip and pick at things, which soft skin wouldn't allow. They also allow us to have a good scratch. The only difference that we know of is that fingernails grow faster than toenails, increasing in length by 0.5mm per week. Both sets of nails grow faster in hot weather than in cold weather. It takes 4 to 6 months for a fingernail to regrow, 12 to 18 months for a slow, old toenail.

Why does a **thumb** become a thumb and not a **toe?**

It seems such a simple question but you have stumbled upon one of the greatest mysteries in developmental biology.

The answer seems to lie in a set of genes discovered in 1983 called homeobox genes. These contain DNA and were first discovered in fruit flies but have now been found in all sorts of creatures including worms, sea urchins, chickens, mice and humans. The homeobox genes are 'switched on' in sequence along the length of the body as it develops, so that at any given point a different set of genes is activated, resulting in a leg, arm, wrist, as appropriate. So, as arms grow out from a higher point along the body axis than legs, a different combination of homeobox genes are switched on in the cells destined to become the arms, from those that will become legs.

JUST WONDERING . . .

Why does **blood** turn a **rust-like** colour when it **dries?**

Rust-coloured is a very good description because blood does contain small amounts of iron which oxidise when they're exposed to air. You could say that, in exactly the same way that iron railings rust, so does our blood.

Why is snot **green?**

Snot, or nasal mucus, is far from being a nuisance. It is a protective fluid secreted from membranes in the nose and we make about a cup full of it every day – more when we have a cold. It is important in protecting the lungs from infections drawn in with our breath, and together with the fine hairs which line the nose, it forms an excellent filtration system.

Snot is only green when it becomes infected. It should be white or clear, but we rarely see it as white because it gets dirty. Dust and particles, such as car exhaust emissions, get trapped in nose hairs, mix with the snot and make it a dark colour. The green colour is due to immune cells called neutrophils. These are the first cells to appear when bacteria start infecting your nasal passages and to do their job require the presence of enzymes, one of which is a ferrous (iron) compound. This is where the colour comes from.

Can **chewing gum** get 'tangled up' in your **intestine?**

This is the sort of thing parents often say to children when they can't stand the sight of them chewing a moment longer. It has absolutely no basis in scientific fact. Chewing gum may adhere nicely to hair, shoes and the bottom of bus seats, but gum passes straight through the stomach and on to the intestines where it has a difficult time sticking to the moist and slippery walls of the gut, and so will pass straight through that as well.

Originally, chewing gum came from the milky white sap of the sapodilla tree found in Mexico and Guatemala where it was collected by farmers and shipped to gum factories. The ancient Greeks chewed gum collected from trees.

Modern gum is a synthetic gum mixed with sugar, corn syrup and flavourings. Over half of a stick of gum consists of sugar.

Why do some of us get **freckles**, and why do they come out in the **sun?**

A freckle, properly called an ephelis, is a circular version of a suntan. Freckles are more common in red or sandy-haired individuals. They are simply due to the sun. If you are genetically prone to freckles and have been exposed to the ultraviolet radiation of sunlight, production of the pigment melanin increases in the pigment cells of your skin. These cells are called the melanocytes which, incidentally, are also responsible for the colour of your hair and eyes. It is a protective mechanism which provides a 'sun-block' layer to stop the potentially harmful rays of the sun doing you damage. Freckles do not form on surfaces that have not been exposed to sunlight.

Freckles usually appear after the age of five and tend to fade somewhat in adults. Apart from avoiding sunlight, there is no known way of preventing them but rest assured that freckles can never do you harm.

What sorts of **bacteria** do you find on your **hands**, and does **washing** get rid of them?

The surface of your skin is a relatively hostile environment for bacteria since it is dry and salty – two conditions which bacteria don't enjoy. Even so, some survive. These include staphylococcus epidermis and acinetobacter calcoaceticus which, in themselves, tend not to be harmful but are somewhat opportunistic and if they spot a chance to make you ill (because you may be poorly already) they will have a go. It has been estimated that there are between 10,000 and 10 million bacteria on each hand.

When you wash your hands, you remove many of the bacteria along with dirt and dead skin, and the soap helps particles to be lifted up off your skin into the water and carried away. But washing your hands doesn't kill the bacteria, it simply removes them.

If you are tempted to use an anti-bacterial soap, you might want to think again because it could contribute to the development of a strain of bacteria resistant to the soap, which would then require yet another antibacterial agent to kill them. Experts say soap and water properly used is fine.

Why is it that when we **bump** our head a **lump** appears?

Our scalp and foreheads have plenty of blood flowing around them, all part of being near the hungry brain. If we do some damage to our heads, such as walking into a door or being hit by a ball, there are plenty of opportunities for bleeding to occur just beneath the surface of the skin. If the bleeding is confined to one small area, as it would be if you'd had a simple knock, then a swelling occurs which, because of its shape, is often called a 'goose egg' – a haematoma would be the correct description.

The immune system is also partly responsible as it sends 'helper cells' to take nourishment to the damaged part of

your head, so they can carry waste cells away. These helper cells clustered around the point on your head where you bumped it are also part of the swelling. The size of the swelling isn't necessarily an indication of the severity of the injury. Even a minor injury to the scalp can cause a large lump. Although the swelling gradually goes away within a few days, discoloration of the skin may persist for one to two weeks. Of course, if you experience other symptoms such as dizziness or blurred vision, you should see a doctor.

Which is the smallest **muscle** in the body, and which the **largest?**

The smallest muscles in the body — and the smallest bones, incidentally — are in the middle ear. They are called the stapedius (which limits the damage done by loud noises) and the tensor tympani (which protects the eardrum). When not acting to protect the ear, these tiny muscles allow the delicate hearing apparatus to move so that we can hear sounds clearly.

Which body muscle is the largest is a bit more difficult. The most powerful muscle is probably the gluteus maximus, the muscle that makes up the bottom and is much used when climbing stairs or getting out of a chair. The longest one is the sartorius, which runs from the hip to the knee, and the largest in surface area is the latissimus dorsi, the broad muscle which covers the back.

How many **veins** are in the human **body?** Do all people have the same **number** of **veins?**

The total length of all the blood vessels in the body is approximately 97,000km (60,000 miles) which is twice the circumference of the Earth at the Equator. However, the majority of these vessels are actually capillaries, not veins, so it's very difficult to give exact numbers. The capillaries allow nutrients and oxygen to pass from the blood into the body tissues, and also remove waste products, such as carbon dioxide, from the tissues by allowing them to pass into the blood. The veins, which are thinner and less flexible than the arteries, take the de-oxygenated blood back to the heart.

Generally speaking, everybody has 34 vein groups, but not everyone will have exactly the same number of veins. If nothing else, the smaller you are the less blood you have, and therefore the fewer blood vessels you will require. However, everybody has veins and arteries, which go to all the parts of the body, and there are at least 34 main veins, and loads more smaller veins connecting with the capillaries.

How long would it take to **count** each **nerve** connection in the human **brain** if you counted at the rate of 1 per **second?**

There are 10 billion nerve cells in the brain, and each of these nerve cells has the potential to connect with 10,000 others, although that is the maximum number of potential connections, so there may well be fewer connections occurring in your brain at this moment.

To work out the total number of connections these two numbers have to be multiplied together. This gives the number of nerve connections in the human brain as 1 x

10^17, which is 1 followed by 17 zeros. To put this number into some kind of context, it has been estimated that there are about 1 x 10^20 stars in the universe. That means there are only 1,000 times more stars in the universe than there are nerve connections in your brain.

So how long is it going to take you to count all these connections? Well, if the rate is 1 per second then it will take you 1 x 10^17 seconds.

What can you tell about your **health** from your **tongue?**

According to ancient Chinese and Indian medicine, your tongue can reveal a huge amount about your state of health. The Chinese even have a tongue map from which they can deduce everything from liver problems to constipation and even anger. A healthy tongue should be pale red with a thin white coat. It should be slightly moist – not too wet nor too dry – and neither flabby nor stiff. If it's cracked then that suggests health problems too.

A hairy tongue is a not uncommon, if unpleasant, occurrence. It is the result of an overgrowth of the normal projections on the tongue which can happen after fevers, antibiotic treatment, or overuse of peroxide-containing mouthwashes. Redness of the tongue may indicate a vitamin deficiency, and a smooth, pale tongue may be a result of an iron deficiency because of the loss of the normal hairy projections.

Don't forget how remarkable your tongue is. It has to work with your lips and teeth to make the sounds we call speech; it moves food around your mouth with the help of cheek muscles; with the help of the nose it gives us the sense of taste, and it helps clean your mouth after food.

If a person **exercises**, does the blood supply to the **brain** increase or **decrease?**

We all know that as we exercise harder, the heart beats faster – you can feel that pumping sensation – and so you would expect the blood supply to the brain to increase. But it doesn't – it stays pretty much the same. The average heart pumps 5 litres of blood around the body every minute. Around 750 millilitres of this goes to the brain and 600 millilitres goes to the muscle(s) working the hardest – say the hamstring. When you exercise, the heart pumps 17 litres of blood around the body – 14,000 millilitres of which goes to the hamstring and, again, 750 millilitres goes to the brain.

How does **oxygen** get from the **air** into your blood **stream?**

Air contains 21 per cent oxygen. When you breathe in, air enters your lungs and the gas passes into the blood in the alveoli of the lungs. The alveoli are tiny air sacs surrounded by a network of blood capillaries. The walls of the alveoli and the capillaries are so thin that the blood and air come into contact, and the oxygen passes through the wall into the bloodstream.

Even cleverer than that, the waste carbon dioxide travels in the reverse direction, from the bloodstream through the thin walls of the alveoli, and back into the stale air which we exhale.

What **happens** when we get **old?**

A very deep and meaningful question this, and one that has troubled many people in the past. Getting old starts early, about the age of 20, but the major changes take place from 40 to 50 onwards. When you are young, the number of new cells which your body produces exceeds those that are dying, and that is how we grow. As we get older, the number of cells that die exceeds those which the body produces. Some organs in the body take a harder knock than others, such as the brain. When there is substantial loss of brain cells, that is where serious problems can begin.

The end result of all of this is that, to some degree, all organs start to function less well as you get older; muscles become weaker, bones become thinner, and mental function deteriorates. Then there's the added problem that every time a cell is replaced and the DNA within the cell is copied, there's a chance that it can be damaged or mutated, in which case the new cell won't function properly. It is also thought that as the working cells grow older, mutations occur so that the proteins produced by those cells are no longer capable of all their original functions.

One of the classic ageing symptoms is a general stiffening of connective tissues which happens in all mammals. Collagen, a fibrous protein found in bones, skin and tendons, is constantly produced early in life. But as we grow older the production of new collagen ceases and the connective tissue consists increasingly of the stiffer, insoluble form – one effect is to make the skin less elastic, which causes wrinkles. Something similar happens in the walls of the blood vessels, and the stiffening of these can lead to higher blood pressure as the blood flow becomes constricted. This in turn requires the heart to work harder, but the poor heart may already be weakened by failing muscles.

It's all part of getting older.

2

LIFE
in the
Wild

Sleepy Bears to Smiling Crocs

Mighty Ants to Half-Dead Worms

Purring Cats to Belly Buttons

Bats to Exploding Seagulls

SLEEPY BEARS TO SMILING CROCS

Do bears **hibernate?**

Hibernation, which is a state of dormancy or inactivity, is used by some bears to adapt to a shortage of food in the winter. But their hibernation is not the same as that of bats, squirrels and a number of other rodents which may break their hibernation to feed, especially if the weather turns warmer. They may also move their hibernation spot if they feel threatened. Bears, on the other hand, enter a deep sleep from which they cannot be roused, accompanied by a dramatic drop in their metabolic rate and body temperature. In some cases, their body temperature drops to near freezing and they enter a state close to suspended animation.

Not all bears hibernate. Bears who live where the winter does not get too cold, and who can find food throughout the coldest season, do not hibernate. Bears that have not put on sufficient fat stores may not hibernate, or do so only for a short time. On the other hand, some bears have been known to be in a state of hibernation for seven or eight months.

Hibernation is a major shut-down of all the bear's systems. They do not eat, drink or defecate; their urea is re-absorbed through the bladder wall, and instead of producing a fatal build-up of nitrogen, it is safely converted into usable amino acids and protein; their body temperature drops several degrees from a normal 98.5, but never below 89; their breathing drops to 8 to 10 breaths per minute from a normal 98; they require only half of their normal oxygen intake; their digestive organs and kidneys shut down almost completely.

Remarkably, there is no permanent loss of muscle functioning despite the fact that they are not using them; there is no loss of bone mass and they do not dehydrate even though they burn 4,000 calories per day just keeping themselves alive. Extraordinary!

Why do **squirrels** have bushy **tails?**

Large tails are useful. For a start, a thick and bushy tail helps a squirrel to balance, enabling it to quickly change the position of its centre of gravity. This is handy when darting through treetops. It can also act as a kind of parachute to break any fall, and just before hitting the ground it can quickly change position to help the squirrel land in a way which will do it the least damage. In addition it acts as a useful blanket which will help preserve body warmth during hibernation.

The tail is also an important part of squirrel communication: tail flicking, for example, is thought to mean 'get off my patch!'

Are the horns on the heads of **giraffes** lightning **conductors?**

Lightning conductors work by allowing the positive charge (which accumulates on the ground under a negatively charged thunder cloud) to leak away into the atmosphere. If the charge is allowed to build up so that there is a large difference between ground and cloud, you get lightning. Flesh and blood is not a terribly good conductor of electrical charge and so horns would not act very effectively as lightning conductors. If a giraffe's horns were made of steel, that would be a different matter.

But if you think of it in terms of evolution, the chances of a giraffe being hit by lightning are so low that even if their horns were some sort of lightning conductor, it is unlikely that would become a dominant characteristic.

How much **dung** does an **elephant** produce?

Watch out, it's a big amount! An elephant produces about 150kg of dung a day (that's about 24 stone). It adds up to 1 tonne of dung a week.

Do **animals** get **addicted**
to substances just like
humans do?

Yes, it seems that animals, like humans, can get addicted to substances that are of no biological value. Rats trained to inject themselves with cocaine continue to do so even when they start having extreme reactions, like seizures. We cannot, of course, assume that rats become addicted for the same reasons as humans, or by the same mechanisms, although the similarity in our brain structures does argue that something similar may be happening.

Not all animal addictions are bad, though. A good example of a useful addiction is the koala bear's love of eucalyptus. Koalas feed exclusively on eucalyptus leaves and will die without them. This addiction is acquired as the koala cub gets used to the eucalyptus in its mother's milk and can't do without it when it becomes an adult. This rather odd addiction has some significant benefits for the bears: eucalyptus leaves contain water which they badly need in a hot climate, and also contain aromatic oils that help keep their fur free of parasites, relax their muscles and keep their blood pressure down.

Many animals in the wild eat plants, fruits and berries (for example, opium poppies and rotting fruits) that contain intoxicating substances.

Is it true that **birds** eat bits out of **alligators'** teeth?

They say it happens, but it's difficult to find anyone who has seen it. Some birds are said to pick food from the gums and teeth of crocodiles in Africa, usually those that are basking peacefully in the sun with their mouths partly open, and it is thought that the Egyptian Plover or the Spur-winged Plover may be responsible. Both these species feed close to basking crocodiles, so it may be possible. The Common Sandpiper also feeds near crocodiles when in Africa during the northern winter.

How **strong** is a crocodile's **bite?**

The muscles that close a crocodile's jaws are very strong – they can crush turtle shells with ease. The large Saltwater Crocodile can crush a pig's skull with the slightest movement of its jaws. So the strength of a crocodile's jaw as it closes is impressive, but the muscles involved in opening the jaws have little strength and a mere rubber band around the snout is enough to prevent a 2m long crocodile from opening its mouth.

Why did the **crocodile** survive, but the **dinosaur** became **extinct?**

Crocodiles are animals which are very flexible in their response to environmental change, whereas dinosaurs evidently weren't. The same applies to turtles, snakes and lizards who could also have died out, but survived because they were able to adapt to changes in their environment. Whether this involved them becoming scavengers or lowering their metabolic rate or living in a hole in the ground doesn't really matter; the fact remains they were able to do this. Dinosaurs couldn't get a hold on the idea of adapting, and that is one of the reasons they died out.

Crocodiles have remained largely the same shape for millions of years, simply because it works so well for them. Did you know that crocodiles don't chew? Along with their food, which they swallow whole, they also take in a few stones, and these stones help to break up the meat and bones in the crocodile's first stomach before it passes to the second for digestion. Crocodiles have more acidic stomachs than any other vertebrae.

MIGHTY ANTS TO
HALF-DEAD WORMS

How **strong** is an **ant?**

Stronger than you think. Ants can carry up to fifty times their own body weight on their backs, and their pincers can grip something 1,400 times their weight.

Where do **ants** go in the **winter?**

Ants live in colonies where all the ants have different roles: there are workers, soldiers, nurses looking after grubs, and of course the Queen who does nothing but lay eggs. She can live up to twenty years.

An ant colony is busy all year round. During the summer and autumn you may see lots of ants scurrying around for food to feed the grubs. When the weather starts to get cold, the entrance to the colony is sealed up and the ants go deep inside where they perform a kind of hibernation, doing only what needs to be done to keep things ticking over. During the winter they survive because they have built up food stores in their body which they can live on until the weather improves. Then, in springtime, the entrance to the colony is unplugged and the ants start being busy workers again.

Why do **woodlice** need the **damp?**

Woodlice fall into a category of animal called Crustacea because they have a hard shell. They evolved from sea-based life forms – as opposed to insects which evolved from land-based life forms – and they are the only crustaceans which have successfully made their homes on land without needing to return to the water to breed. Their skin is permeable, which means it lets water in and out. If they did not live in damp places they would dry out and die. That's why they like damp places. Interesting fact – woodlice breathe through holes in their back legs.

Where does the **bluebottle** fly sleep at **night?**

Flies don't sleep in the sense that they close their eyes for a spell and wake up refreshed. For a start, they don't have eyelids so their eyes are open all the time.

But they do rest during both the day and the night in what are called quiescent periods where their metabolism slows, including their breathing and heart rates, and they start to use far less energy. Apart from having no eyelids, they need to remain vigilant to ensure they are not eaten by predators, so they don't fall into what we would recognise as sleep. But they do fall easily into a 'quiet' state when it gets dark, and they will stay that way till it gets light again.

How do you sex a **caterpillar?**

There's no need to tell the sex of a caterpillar because they don't have any. They are completely sexless – they are simply immature forms of their own species. You can think of them as being simply bags of stuff which eat a lot. They have no sexual organs, either internally or externally, and the only way to determine their sex is to analyse their DNA.

Why don't **butterflies** get stung by **nettles?**

It is true that some species of butterfly (and caterpillar) can feed quite happily on stinging nettles without showing any signs of the reaction which we would call stinging.

The reason that humans get stung by nettles is because the nettle produces two chemicals; acetyl choline which gives the stinging sensation, and histamine which produces the itch. Our skins are sensitive to both these chemicals. Butterflies don't have a skin like we do, but a tough armour-like exoskeleton, and they are not affected by these chemicals in the same way.

Is there anything **good** about **cockroaches?**

Lots! Cockroaches deserve respect. They are among the oldest creatures on Earth, and some rocks said to be over 200 million years old contain cockroach fossils remarkably similar to the creatures which are such a pest today.

They are curious creatures: for a start they bleed white blood, their skeleton is on the outside of their body and, as they grow, they shed their external skeleton several times a year. A cockroach that has shed its skin is white with black eyes. After eight hours it will regain its regular shell colouring.

Amazingly, a cockroach can live for a week without its head. A beheaded roach only dies because it has lost its mouth and can't drink water, so it dies of thirst. They can also hold their breath for 40 minutes.

They don't like daylight, favour living in warm nooks and crannies and will eat pretty much anything during their preferred eating hours which are just after dark and through the night. They are a particular pest in hospitals and kitchens, where they can carry a range of nasty diseases, and they are rumoured to feed off human eyebrows, given a chance – but this may be one of those tall tales. Disgusting though they are, and despite our recent attempts to limit them, they have survived on this planet 100 times longer than we have. That's why respect is due.

Why do the bites of **horseflies** hurt so much, and **mosquitoes'** don't?

Horseflies don't suck blood like mosquitoes. The mouth parts of mosquitoes are built for piercing and form a sharp, prominent proboscis which extends forward from the head. The fleshy part of the proboscis houses the stylets which are thrust into the skin like a hypodermic needle. Because of the fineness of that needle, a mosquito can suck away at you for some time before you realise it.

Horseflies are different. The mouths of horseflies are more blade-like, intended for cutting through skin. The bite of the horsefly is more painful because the cutting parts of their mouths inflicts much more damage to the skin than the stab of a very fine needle.

Why do **cats'** and **rats'** eyes **glow** in the **dark?**

They don't really glow. There is a layer at the back of the eye called the tapetum, which is formed of fibres which are reflective, and it is these which reflect the light back out of the eye like a mirror. The tapetum, which is behind the retina, reflects the light so well that the eyes stand out from the dark background and appear to glow. One of the results is that cats can see far better in the dark because of the reflective, light-enhancing properties of the tapetum — they can operate with only one sixth the illumination that the human eye requires.

Why do **midges** always **bite** me, and leave other people **alone?**

It's because you smell. We all smell differently, and insects can detect these differences and express their preferences by feeding on some people and not on others. Some chemicals which the body produces, and which are known to attract insects such as mosquitoes, are carbon dioxide — which we all exude every time we breathe out — lactic acid and octenol. Insect repellents work by masking odours that might attract insects, or by creating smells that they find repulsive.

The midge (*Culicoides Impunctatus*) is commonly found in Scotland, but exists in many other parts of the world – they are called sandflies in Australia. It is only female midges which bite – they do this in order to suck your blood to feed themselves before laying eggs. Unlike mosquitoes, midges do not carry diseases.

Why do **snails** and **slugs** produce a **slimy mucus** and how do they **do it?**

The real question here is how does an animal with one foot walk on glue? The mucus they produce is not only a lubricant, it is a glue as well – properly described as 'visco-elastic'. By exerting different shear forces on the mucus with its foot, the mollusc can change the properties of the mucus, allowing it to slide gracefully over a rough surface one moment, and stick firmly to a vertical surface the next.

The mucus comes from several very large glands in the foot which are filled with cells called goblet cells. The goblet cells manufacture glycoproteins, which are protein molecules with a large number of sugar molecules attached. We have them in our body too – the mucus in our throats and noses is a glycoprotein. To watch a mollusc in action, persuade one to crawl over a sheet of glass and watch it from underneath. You'll see it sliding over the mucus produced in the centre of the foot, while the muscles on the edges of the foot ripple and propel it forwards.

Can **snails talk**
to each other?

It appears that snails and slugs have two ways in which they communicate. On their head and face are two pairs of tentacles; the top ones are longer and have simple eyes at the end. This doesn't mean they have much vision but if something large came at them, like a hungry bird, then they'd see it coming. The lower pair of tentacles are used for touch and to receive chemical signals, and this is what happens when two snails who fancy each other meet. They send 'loving' chemical messages between themselves, although remember that it is not like a loving human relationship because snails are hermaphrodites, which means they are both male and female at the same time.

When a snail is exploring its environment, it will wave all its tentacles through the air to pick up as much information as

possible; but at the same time it will leave behind messages about itself in its slimy, mucus trail, which other snails can detect and 'read'.

How much **food** does a **whale** eat in a **day?**

If it was a blue whale it might eat up to 4 tonnes of krill every day – krill are small, shrimp-like creatures each weighing about 1 gramme – so the whale would be eating at least 4 million a day. Because whales are not diet-obsessed, no one seems to have worked out the caloric value of a krill, but 100g of shrimp contains 106kcal, and that would be pretty close. So, if we use that value, then the blue whale takes in 4 million calories per day. If you think that sounds rather fattening, it is – a whale can gain as much as 770kg per week. However, it only feeds for 4 months of every year. For the rest of the time, the whale fasts while migrating.

Why do **snakes** survive on a couple of **meals** a year, while **sheep** eat all the time?

Snakes have a very low metabolic rate, which means they convert food into energy very slowly. Instead of burning food to keep themselves warm, they gain heat from the environment to heat up their bodies. They also obtain prey by ambush, which requires the least amount of energy. Reptiles are very efficient at digesting what they have eaten, and have very acidic stomachs which can even digest bone and shell. Crocodiles, for example, have vast fat reserves to keep them going from meal to meal.

Sheep, on the other hand, are very inefficient at digesting the cellulose found in grass, so most of their food passes straight through their gut undigested. This means they have to eat vast quantities to obtain the nutrients they require to stay alive.

The two have evolved in these different ways because of the food that is available to them. Snakes and crocodiles have an intermittent food supply, often dependent on migrating animals which might not come their way very often, whereas sheep have a more consistent food supply and can grab a snack whenever they feel like it.

Can there be **more** than two **sexes?**

The simple answer is no, although there are some species which have males, females and hermaphrodites – organisms that can be both male and female at the same time. But there is no third sex.

Why are there only two? The whole point of sex is to increase the genetic variation in the population to help the species survive, so you might think more sexes would mean more variation. But having more sexes also increases the problems caused when different cells combine.

The bulk of a cell is cytoplasm, a jelly-like material which completely surrounds the nucleus of cells. The cytoplasm acts defensively, so if two cells attempt to combine the cytoplasm ensures that no combination takes place. But sperm have no cytoplasm to prevent this happening. If there were more than two sexes there would still have to be some who produced sex cells with cytoplasm and some without, and they would still only successfully mate with the other type, so the variation would come back to two sexes.

Male seahorses are the ones that 'get **pregnant**'. Do any other **animals** do that?

It is true that it is the male seahorse which becomes pregnant. He carries a pouch below the chest area, and it is into this pouch that the female lays the orange-coloured eggs. The male then sways around a bit until all the eggs are evenly distributed, and within three weeks there could be anywhere from 50 to 1,500 baby seahorses born.

74

Pipefish females also lay their eggs in their partners' pouches, and it is the males that carry the offspring. Also, the female golden egg-laying beetle dumps her eggs onto males that are mating with other females, so the males do all the carrying. This is a pretty risky job as the eggs are bright yellow which makes the males very easy to see and much more likely to be eaten.

How many **eyes** does an **earthworm** have?

None. Earthworms have a very basic body structure and physiology. They don't have any eyes or ears either, although their entire body (particularly the upper surface) is incredibly sensitive, especially to light. They can also sense the vibrations of a mole digging, and many earthworms will rise to the surface if they sense one is tunnelling close to them. The reason for a worm's sensitivity is that it has a nervous system which runs the entire length of its body, and a large collection of nerves which bundle together in the head region – these are collectively known as the 'ganglion'.

Earthworms are hugely beneficial to the soil in which they live. They maintain channels through which air and water flow, and they drag organic matter deposited on the surface deep into the soil. They do this by shredding the leaf and partially digesting it, then it passes through their bodies emerging as highly fertile worm casts. Charles Darwin, the great naturalist, once worked out that if you took just one acre of soil and collected all the worm casts produced in ten years, you would be able to spread them on that acre of soil to a depth of two inches.

If you cut a **worm** in half will both **halves live?**

No, both halves will not live. The end with the head will survive if the cut is made behind the saddle, which is the fat bit of the worm where the major organs are located, because, in theory, the worm will be able to grow a new anus. If the cut is made in front of the saddle, the worm will certainly die.

PURRING CATS TO BELLY BUTTONS

How do cats **purr?**

In domestic cats it is the vibration of an elastic ligament linking the clavicle bone to the throat, and the purr is created during both inhaling and exhaling. In their larger cousins, such as lions, things are slightly different and the purring only occurs on the out-breath. Cats never stop purring – they just control the volume.

Purring is usually associated with contentment, such as stroking or cuddling. But purring can equally be a sign of stress, which is why cats purr loudly while being examined by a vet. Some research has suggested that purring can improve bone density, and might be part of a healing process, and that the constant purring is part of an ongoing maintenance programme which requires little energy. Certainly, vets will tell you that a cat's broken bones mend more quickly than in other animals – possibly one of the reasons for a cat being said to have 'nine lives'.

Cats have **whiskers** to help them judge **space**. So, if a cat is **pregnant**, does it grow longer **whiskers?**

The length of a cat's whiskers are set genetically, and aren't adjustable. If a cat grows fatter than normal, for whatever reason, its whiskers might be too short and in theory the cat could get stuck when passing through small spaces.

But in the case of pregnant cats it's not too much of a problem. They are only pregnant for 9 weeks, which doesn't give them a great deal of time to get stuck in doorways. The other thing to remember with cats, whether pregnant or simply fat, is that they are the sort of creature who would rather be lying in front of the fire, which is where they would prefer to be stuck.

Why do cats **sniff** each others' **noses** when they first **meet?**

It is probably something in the saliva, or on the breath, that is a unique identifier of individual cats. Rats and mice do the same and you have to remember that these, and many other mammals, are primarily nocturnal and need to be able to recognise friend or foe in the dark. Of course, cats also sniff each other at the opposite end to the nose, and this gives them clues to sex, maturity and social status.

The sense of smell in cats (and dogs) is highly developed and said to be over ten times stronger than ours. Beagles have been used to warn diabetics if their sugar levels are rising – the dog smells it and barks a warning.

Why don't **cats** go **grey** when they get **old?**

The colour of a cat's fur is due to the presence of melanin in the hairs. Melanin is usually black, although it can be converted to a chocolate brown colour. There is also another form of melanin, called phaeomelanin, which is an orange or yellow colour. Sometimes the hairs do not contain any pigment, then they are white.

All cat coat patterns are made up of different combinations of these colours. The pigments melanin and phaeomelanin are produced by special cells called melanocytes which are present in the hair follicle, so that the pigment is incorporated into the hair as it grows. The pattern and colour of any cat's coat depends on which colour pigment is produced and when, and this is controlled by the cat's genes. In some cats, the hairs are the same colour along their entire length, but in other cats the colour of the hairs varies along the length. Sometimes the hair is yellow/brown for most of its length but with a black tip, or sometimes the hair has black bands all the way along it. This happens because the melanocytes produce different kinds of melanin at different stages in the hair's growth.

With increase in age cats tend to lose some of their colour but they do not go grey all over, instead their fur tends to be duller and the colour not as strong. Why cats do not go grey all over is not clear, although in general they grow old far more gracefully than we humans do. Perhaps this is due to their diet or the amount of sleep they get.

Do **cats** have **belly buttons** and, if so, where?

Cats do indeed have a belly button and you'll find it pretty much in the same place as on humans – just below the rib cage. But don't expect it to look much like a human belly button. If you've got a willing cat and are prepared to dig around a bit because it's covered by fur, you'll find it looks like a scar.

Belly buttons are the remnants of the umbilical cord by which the foetus is attached to its mother and through which flows blood between foetus and placenta. In the case of cats it is the mother's first job after the kitten's birth to cut the cord with her teeth.

BATS TO EXPLODING SEAGULLS

Why do **bats** hang **upside down?**

Their leg arrangements are not the same as in birds. The thin membrane of a bat's wing stretches from its elongated fingers to its legs, and is then attached to the sides of its body. Most bats also have a large membrane connecting both legs, which acts like a pouch and is used to capture insects. This means the legs of a bat are not free to do their own thing, but are 'tangled' in wing and tail membrane.

This is why bats were not able to develop the many different kinds of legs you find on birds – long legs in storks, short legs in ducks. For bats, it appears that the easiest way to stow away the membrane is to hang upside down. In other words, it's the most comfortable position for them to be in.

Why doesn't the **blood** in **bats** run to their **heads** when they're hanging **upside down?**

The circulation of blood in your body adapts quite readily to whatever position you are in, and so does the blood in bats. If you stand on your head, your heart, arteries, capillaries and veins all work to move the blood through the circulatory system in the necessary direction, no matter how your body is positioned.

In the case of bats, their hearts are quite large in relation to their body size, compared to other mammals. They also have high stroke volumes, which is the amount of blood pumped by the heart in one beat. So the combination of large hearts, high heart rates, and high stroke volumes assists their upside-down resting position. In addition there are anti-back flow valves in the heart, and special modifications of the arteries, veins and capillaries in the flight membranes.

Why don't **bats** make **sounds** we can **hear?**

Bats make sounds all the time but our ears aren't sophisticated enough to hear them. Bats produce noises at a very high pitch – ultrasound. Like dolphins and whales, bats use sound to locate their prey and to work out where they are and where they want to go. It's called echolocation. The sounds the bats make bounces off objects like buildings, trees, animals etc. Bats listen to this echo and they can then determine how far away the object is and what its shape is.

Why do some **birds** hop and others **walk?**

Ground-dwelling birds, such as pheasants, tend to walk; songbirds which live in trees tend to hop as they travel from branch to branch. Parrots often walk along branches, and house sparrows hop when they come to the ground, while palm warblers walk on the ground and some songbirds, such as American robins and European blackbirds, may both walk and hop. Some birds with small feet, such as swifts, hummingbirds, bee eaters, and many hornbills, use their feet only for perching and rarely walk at all. Other birds with robust feet, such as guinea fowl, do most of their moving about on foot. Whether birds hop or walk appears to depend on how they have adapted to live in a particular environment – the ground is good for walking, while moving from branch to branch is best done by hopping.

Do **birds** have **earwax?**

No, birds do not have earwax. Birds have only a very short, or in some cases virtually no, external ear canal, and the tympanic membrane – the ear-drum – is stretched across the opening. The purpose of earwax is to protect the membranes in the ear canal, but since they don't have one they have no use for the stuff.

Do **birds** have **bladders?**

No, they don't. Most of the liquid gets absorbed and recycled, adding to what are mostly solids which pass through the bird, and emerge as faeces. There is only one hole for stuff to come out of.

How did **birds** get **feathers?**

Scientists are still arguing this one, and theories go in and out of fashion, but the most widely held view is that feathered dinosaurs, such as the famous *Archaeopteryx* fossil – said to

be 150 million years old and considered by many to be the first bird – evolved their feathers from scales initially used for insulation. These were ground-dwelling creatures, and it is thought that when they evolved feathered arms, like wings, they began to use these to catch insects. Some theorists think that the random flapping involved in this food-catching eventually led to the evolution of flight. It sounds unlikely, but is the accepted viewpoint at the moment.

But there's another theory; feathers evolved in tree-dwelling dinosaurs in order to help them break their falls when leaping from tree to tree. You can believe that one, if you like. There is no real answer. The *Archaeopteryx* fossil is by no means the most important fossil. Other dinosaurs, both before and after it, have possessed feathers and may have exhibited powered flight.

Why don't birds' **eggs crack** when they're **laid?**

The shell of a hen's egg takes up to 16 hours to develop inside the bird's body and is formed from crystals of calcite, which is a type of calcium carbonate reinforced with protein fibres. It is completely hard when laid, just as you find it in the egg box. The shell of the egg is quite strong, and well able to withstand the stresses of passing down the oviduct, which is a muscular tube which contracts to move the egg along. If you were to put an egg in your hand and then close your fingers around it, you'd find it would take a surprising amount of force to crack it. The egg even survives the forces of being expelled, and then sat upon by the hen. The oviduct is also capable of stretching to accommodate the bulk of the egg, so that the egg does not become crushed, or crack.

How can you change the **colour** of a hen's **eggshell?**

Short of painting it, you can't change the colour of an egg. The colour is a matter of genetics. Contrary to popular belief, what you feed a hen has no effect on the colour of the shell, and the colour has no nutritional significance. Some people think brown eggs are tastier than white ones, and vice-versa, but there's no truth in it at all. If you want white eggs, for example, you have to go to a hen that always lays white eggs, such as a White Leghorn. It is the breed of hen that determines the colour of the shell.

Why don't we see **baby pigeons?**

Pigeons don't leave their nest until they are about 35 days old and by then they must have fully-formed feathers to be able to fly. In other words, by the time they are ready to leave the nest they look like fully-grown pigeons. So we don't see them when they are very young because they are still in their nests.

Where do **ducks** have their **ears?**

Ducks have their ears in the normal place, on either side of the head. But they don't have ears that stick out like we do. Instead, the opening of the ear canal is covered with a layer of special protective feathers, which stop the water getting in, which is why we can't see them.

Why can't **seagulls** land in **trees?**

There is nothing in the anatomy of a seagull that prevents them from landing in trees. In fact, there are some gulls that nest only in trees; in particular the Bonaparte's Gull nests in coniferous trees and builds its nest at about 15ft above the ground on horizontal branches. But the more common Herring Gull can nest in many different habitats including cliffs, building ledges and trees.

So, gulls can land in trees but whether or not they do so depends on which habitat they prefer. Some will not land in trees because they are specialists that only nest on cliffs.

87

If you gave seagulls
Alka-Seltzer,
would they **explode?**

I suppose the thought behind the question is this: because a seagull's stomach is effectively a large bath of acid, the arrival of the alkaline Alka-Seltzer would cause an instant reaction in which large and possibly destructive quantities of carbon dioxide would be given off. There have also been rumours that you shouldn't throw rice after a wedding as the grains will expand in the birds' stomachs causing them to explode.

The truth is the seagulls wouldn't explode if they ate Alka-Seltzer, and pigeons don't explode when they eat rice. Think about it. If expanding rice caused birds to explode, there would be no birds in China. Many migrating ducks and geese depend on winter-flooded rice fields each year to fatten up and build strength for their return trek to northern nesting grounds.

When it comes to seagulls, one ornithologist told us: 'I can say with 100 per cent confidence the gull would not blow up. Even though a seagull's stomach is small, the gull's digestive system would probably process the Alka-Seltzer just as a human's would. Even if the gull ate the Alka-Seltzer instead of drinking it dissolved in a glass of water, it would probably just dissolve slowly in the stomach.'

3

SCIENCE
All Around

Fizzing Bubbles to Falling Bubbles

Rising Heat to Maple Syrup

Wet Windows to Bending Rainbows

AC to DC

Tangled Bedding to Tasty Toast

FIZZING BUBBLES TO FALLING BUBBLES

When you shake a **bottle** of **fizzy drink**, why does it **fizz** when you **open** it?

The gas that makes drinks fizzy is carbon dioxide, as it is dissolved in the drink like sugar in a cup of tea. However, while the bottle is on the shelf it gradually leaks out and collects in the space between the bottle cap and the drink — called the head space. Usually, when you open a bottle of drink, this gas can escape easily and you only hear a slight psssht.

But when you shake a bottle, the carbon dioxide contained in the head space gets mixed up in the drink and forms pockets of carbon dioxide within the liquid which can't easily escape when you open the bottle. Instead, these pockets rise rapidly to the top of the bottle dragging some of the drink with them. It then flows out over the top of the bottle and makes a fine old mess.

Why does a drink **fizz up** when you drop an **ice cube** in it?

It is very hard for gas bubbles to form in a liquid — they usually need a rough or porous surface on which to form. When chemists are boiling liquids in a glass beaker, particularly liquids other than water, they usually add a few porous clay chips to help bubbles begin to form as the liquid comes to a boil. If they don't, the liquid warms to several degrees above the boiling point without any bubbles forming, and then it 'bumps' – a sort of eruption which could be dangerous.

Once you've opened a bottle of fizzy drink, the dissolved carbon dioxide is trying to turn into bubbles of carbon dioxide gas and escape. But it needs help in the form of a rough surface on which the bubbles can form. Ice cubes are often rough and these do the job nicely. But how much fizz you get when you drop an ice cube into a drink depends on how the ice cubes were made. If the ice cubes are rinsed, the surface layer melts and then refreezes to give a smooth finish and gas bubbles will have a much harder time forming on this smooth surface. For a big fizz, use rough ice.

Once you open a bottle of **fizzy drink**, is it better to **squeeze** the air out or leave the **air in** when you put the **lid back** on?

If you squeeze the air out, carbon dioxide leaves the drink to fill the space so you lose the fizz. If you leave the air in, the carbon dioxide in the air and the carbon dioxide in the drink reaches an equilibrium so the fizz remains in the drink. Therefore it is better to keep the air in the bottle.

Why does a **cold fizzy** drink make your **eyes water?**

Blame the bubbles and the cold. The tiny bubbles have irritated your sinuses, which are the tubes in your head which get blocked when you have a cold. Anything that irritates the sinuses can make your eyes water.

The coldness of the drink also causes the blood vessels in your head to constrict, making it harder for the blood to get through them. This narrowing can also make your eyes water, and give you a sharp headache for a couple of minutes.

Can dangling a **silver spoon** into an open bottle of **champagne** stop it going **flat?**

People in the wine industry will tell you that this really does work. Scientists say there's no truth in it at all. The wine experts say that if champagne is kept cool, it will stay fizzy for twenty-four hours or so after it has been opened because carbon dioxide (the fizz) is more soluble in a cold mixture of alcohol and water than in sugar and water-based soft drinks, which tend to go flat more quickly. So with or without the silver spoon, champagne will hold its fizz. Silver spoons have nothing to do with it.

Here's a theory: it might be possible that the people who are sober enough to remember to put a silver spoon in the bottle, are also sober enough to remember to put it in the fridge.

Why do the **bubbles** in **Guinness** seem to go **down** the pint?

The bubbles in a freshly poured glass of Guinness do appear to be falling to the bottom of the glass, yet at the same time the creamy head is getting larger. For a full explanation, we must give credit to researchers Alexander and Zare from Edinburgh and Stamford universities who dedicated much of their time to staring at pints of Guinness, trying to solve this conundrum. Here is their solution.

Imagine the pint freshly poured and beginning to settle. The bubbles start to rise, as in all fizzy drinks, but those in contact with the glass are slowed down by the drag created where the bubbles and the glass meet. But the bubbles in the centre of the glass can rise freely and faster, and as this flow meets the head, the bubbles and beer spread out to meet the edges of the glass where they are deflected downwards. As the liquid flows down the outside of the glass, it takes the bubbles with it, which is why the bubbles 'float' downwards. This circulation continues for some time, but eventually enough bubbles are deposited in the head and the process runs out of energy and your pint has settled.

This effect doesn't only happen in Guinness but the pale bubbles are easier to see against the dark liquid. Also, the bubbles in most drinks consist of carbon dioxide but in Guinness the gas is nitrogen, which doesn't dissolve as well as carbon dioxide, and so isn't released with the same fury. It makes for a much more tranquil bubble which takes its time to move around the glass.

Why are **bubbles round?**

We need to think about surface tension, which is kind of custard-skin effect. If you look closely at a pond you may see little insects walking about on its surface as if they were on a sheet of rubber – that's because of the surface tension. If you try breaking the surface of a bowl of water with a flimsy bit of paper you'll find the paper will curl before it finally penetrates the surface – again, surface tension.

Surface tension is due to the links between the molecules which make up the water. Let us imagine that the molecules making up the water have six links, or bonds, with neighbouring molecules; one each for the molecule above it, below it, to the left of it, to the right of it, in front of it and behind it. The molecules at the surface of a pond have no neighbours above them, so the spare link goes to a neighbouring molecule on the surface. This means that every molecule on the surface has an extra bond pulling it to another surface molecule. These extra forces make the surface molecules the most tightly bound in all the liquid and that's what creates the custard-skin effect. This effect occurs whenever water meets air, or any other gas.

It also occurs within a bubble. When a bubble fills with gas, it tries to hold it in the most stable structure it can within the smallest surface area. The shape it naturally falls into is a sphere because it has the smallest surface area compared to its volume, and also requires the least amount of energy to create it.

Bubbles that appear in the air are slightly different, although it is still surface tension which holds them together. In soap bubbles, a balance has been achieved between the surface tension of the film of soap, which is trying to make the bubble shrink, and the outwards pressure of the air which is trying to explode it. That's why blowing a bubble calls for fine judgement – it either bursts or collapses if that balance isn't quite right.

Why do you have to add **soap** to water before you can **blow bubbles?**

The molecules in water have a strong attraction. If you try to blow bubbles in pure water, the molecules would never move far enough apart to create the thin film of water that you need.

But if you add soap, the surface tension (created by the attraction between molecules as explained above) decreases hugely and it is possible to create the thin film from which a bubble can be blown. But there is still surface tension, although much reduced, and it is this which gives the bubble its shape because the surface tension is always attempting to pull the film into the smallest possible shape, which is a sphere. A sphere is a shape which has the smallest surface area compared to its volume.

How **big** can a **soap bubble** be?

The surface tension forces are the ones which keep the bubble in shape, but the force of gravity is also acting on the bubble and attempting to pull down the top of it. Once the gravitational forces exceed the surface tension forces, the bubble will collapse. The same effect limits the size of a water drop that falls from a tap – eventually the weight of the water is simply too great for the surface tension to hold it up.

However, since the layer of water in a bubble is very thin and therefore very light, the effect of gravity is not very big and the bubble is more likely to break because the surface tension is disturbed by bumps or the wind. Another thing which helps a bubble to burst is the rapid evaporation of the thin film of water.

Can you blow **soap bubbles** in **space?**

Think about what happens when you blow a soap bubble. You are pressing air against a soapy film, increasing the pressure on it until the film starts to expand and forms a bubble. Even blowing a soap bubble requires some effort because you are having to blow against the pressure of the surrounding air in order to get the bubble started. At the moment the bubble leaves the straw, though, the pressure on the inside and outside of the bubble are the same, otherwise the bubble would explode or implode depending on whether the internal pressure was greater or smaller.

The reason bubbles have a short life is because sooner or later they will tend to rise or fall. If they rise, the external air pressure becomes less, because there are fewer air molecules pressing against them, and the bubble explodes. If the bubble falls, the external air pressure increases and the bubble implodes.

But the more likely reason for your bubble to burst is because there is a small amount of water trapped in the film, and this tends to drain to the bottom of the bubble under gravity. Eventually this weight of water becomes too great for the bubble's surface tension to support, and the film begins to split, and the difference in pressure makes the split grow even larger.

In space, of course, there is no air pressure because there is no air. As soon as you tried to blow down a straw, the soapy film would burst because there would be nothing for you to blow against. So there can be no bubble-blowing in space.

Which of these would hit the **floor** first if you **dropped** them from the same height: a **bubble**, an iron ball or a **wooden ball?**

In air the answer is quite simple: the iron ball would hit the floor just before the wooden one, and a long time before the bubble. This is not because of any difference in weight, or anything to do with gravity, but because lighter objects are more affected by the air through which they are moving. The bubble is clearly wafted along by the air and can get caught up in any slight circulations, so may not reach the ground for some time. Because the wooden ball is lighter, in theory it will be more influenced by the atmosphere than the iron ball, but the difference will be very small.

In a vacuum the situation is entirely different. First of all the bubble couldn't exist as the air pressure inside the bubble would make it explode (see above). Secondly, there is no air resistance to affect either the wooden or iron ball, so both would hit the floor together. If you don't believe me, take a look at the pictures sent back from the Apollo moon mission where they dropped a hammer and a feather. Both hit the floor at the same time.

RISING HEAT TO MAPLE SYRUP

How does **heat** make things **rise?**

If you make any part of a liquid (or a gas) hotter than the liquid around it, it will rise. This is because everything is made of atoms and when you heat things up you are giving those atoms more energy – the result is that the density is reduced. This is because when the atoms become more energised, they rattle about more, and start to take up more volume. Imagine you and your mates are jiggling about in the gym – the more you move, the more room you take up. It's the same with atoms. So although the atoms have the same mass, the volume goes up and the density goes down. Things that are less dense rise.

If you heat a **tin tray** until it is very **hot** and put some **water** onto it, the water forms small **blobs** that **skim** around very **quickly**. Why?

The temperature of the tin plate needs to be much greater than 100°C – the temperature at which water boils – for this to happen. The surface of any water drop that comes into contact with a very hot plate will almost instantly evaporate, and this evaporating water forms a cushion of air and steam that the blobs of water can float around on. As they float around, more and more of the water evaporates to replace the cushion of air and steam that is gradually lost. This creates a large amount of turbulence underneath the blob of water and it is this which sends it speeding all over the tin plate.

Why do you **blow** on a **match** to put it out, but blow on a **fire** to make it **burn?**

It all depends on the strength of the blow. Blow too hard for too long at a fire and it will go out. Blow too lightly on a match and you won't blow it out.

For anything to burn it needs oxygen, heat and fuel. The heat gets the burning going, but you need an initial input of energy to start the process off. With a fire you can sometimes do that by blowing on it because you are adding more oxygen to the system and helping to spread the flames. However, if you blow too hard, the fire will go out because you will have cooled it too much.

When you blow on a match, it goes out because the blowing cools down the area of the match that is on fire and so the fire goes out. So it can work both ways.

When you look at a **match**, why does it burn two **different colours?**

Flames have different colours depending on the amount of heat energy being released. Although we often associate the colour red with heat, this colour actually relates to the least amount of heat energy. If we imagine a blacksmith making a horseshoe, he puts the metal into a furnace. As it begins to get hot it glows red and then turns more orangey-yellow. If he were able to heat it up enough, you could get the horse shoe to glow white-hot. The flame of a candle is generally yellow, while a gas cooker flame tends to be blue – the blue flame is hotter than the yellow one.

A match has a range of heats, and so displays a range of colours. Closest to the wooden stick the flame will tend to be blue, which is the hottest part; while the further you are away from the stick the redder it gets because it is cooler.

If **black** things absorb **heat**
better than white,
why do people in **hot countries**
sometimes wear
black clothes?

There's a good scientific explanation for this, but it may not be the real reason. It is true that black clothes worn by people in hot countries do get hotter than if they wore

white. The result is that the clothes become much hotter than the body and this results in a large difference in temperature between the two. Temperature differences cause air to move, creating a breeze, and this breeze cools the body.

However, it is unlikely to be the real reason why people wear black. It is much more likely worn for cultural reasons.

When you put a cup of **water** and a cup of **maple syrup** in the **microwave** for the same length of time, why does the **maple syrup** get **hotter?**

Some materials need more energy than others in order to raise their temperature. The amount of energy needed to achieve a certain temperature change is a measure of the material's thermal capacity. If maple syrup has a lower thermal capacity than water, then it would heat up faster as it would need less energy to raise the temperature. We don't know the thermal capacity of maple syrup – no one seems to have bothered to measure it – but we have a sneaking suspicion that it might well be higher than that of water.

Another factor is the 'radiation density' of the material. A microwave works by emitting high-energy waves inside the microwave oven. These waves bounce off the walls and hit the food. Some of these waves will be reflected back from the food, but others will pass through, activating the molecules and causing the food to warm. Water is pretty absorbent of microwaves, but not amazingly so. If maple syrup was much more absorbent than water then it would absorb more energy from the microwaves in a shorter time and so heat up quicker than the water.

A final factor is that maple syrup has a lower vapour pressure and boils at a much higher temperature. Water cannot be heated beyond 100°C but maple syrup could probably be heated to around 200°C and thus is able to get hotter than the water.

WET WINDOWS TO BENDING RAINBOWS

When our car is in the **car port**, why doesn't **dew** form on the **windows**, even though it's **open** on all sides?

It's all due to the roof of the car port. During the day, it acts as a sunshade and the shelter it provides prevents excessive build-up of heat. During the night, it reverses its role and prevents rapid heat loss – a bit like a blanket on a bed.

During the day, the air underneath the roof isn't as hot as the air around it and doesn't collect as much moisture – hotter air can hold more moisture than cooler air. Then, at night, when the outside air starts to cool down, the air under the car port stays at a relatively higher temperature so the moisture in it does not condense, no dew forms, and your car stays dry.

When you're **flying** high in an aeroplane, you can often see **water droplets** in the air. Since the outside **temperature** is way below **zero**, why don't they turn to **ice?**

Water can exist as a liquid below 0°C and there are two main reasons why. Impurities in the water reduce the freezing temperature. This is why salt is put on the roads in winter to help stop ice forming because salt can reduce the freezing temperature to as low as −13°C, depending on how much salt is put down.

But the most likely reason for those high-altitude droplets not freezing is because, for water to freeze, the molecules must form the correct structure, usually around some kind of seed, which may be as small as a speck of dirt. In a large expanse of water, like a cup or bowl, you are more likely to find one, so the water readily freezes at 0°C. But at high altitude, there may be no polluting seed to start off the process, so liquid water drops can exist in the atmosphere at temperatures as low as −40°C.

Can you **shake** something **apart** using **sound?**

You can make anything vibrate – soldiers marching across a bridge, an opera singer hitting a particular note, or a diver bouncing on a spring board. You can also make things vibrate at their 'natural' frequency. So if a springboard has a natural frequency of say 1 Hz – one bounce a second – then jumping up and down on that board at half a bounce a second will not produce as big a bounce as jumping on it at a rate of once a second. When a system is forced to vibrate at its natural frequency, it is said to be resonating with the thing forcing it to vibrate.

It has been said that the great Caruso could shatter a glass with his voice alone (although after his death his wife denied he had ever done it). But for such a shattering experience to take place, the great singer would not only have to have hit the resonant note perfectly, but also with great force. To make something break just by making it vibrate, you need to shake it at its resonant frequency to produce the biggest vibrations. But this won't always work since objects can be held together by considerable internal forces which are stronger than the resonant force trying to break them.

Why does a sheet of **glass** sound different to a piece of **wood** when they **break?**

When a material breaks, some of the energy applied causes the broken fragments to vibrate and emit sound. Different materials have different degrees of 'springiness' and so the various harmonics generated will be dampened to different extents, resulting in a different sound.

Why is glass **transparent?**

When a photon of light energy strikes one of the atoms which make up the molecules of the glass, it excites the electrons within that atom. But not all electrons are in the same orbit; they revolve around the nucleus at different levels, only changing levels when they receive enough energy to make the jump.

With transparent materials, these levels are so far apart that light doesn't have enough energy to excite the electrons, and so the light passes straight through without interacting with the material. That's why we can see through glass and water. But transparency is related to the wavelength of the light, and although visible light is clearly visible through glass, at the ultraviolet end of the spectrum glass will become less transparent since the UV rays do have the energy to make the electrons change orbit.

Is it true that **old** windows are **thicker** at the **bottom** than they are at the **top** because the **glass** has **sunk** to the bottom?

Glass does not flow, although it is technically called a supercooled liquid. This is because it has no grain structure and is perfectly elastic up to the moment it fractures. If you look at very old windows where the panes of glass are only 6 or so inches across, you can often see that certain bits of glass are thicker than other bits. Sometimes they occur at the bottom, but sometimes at the side or even the top. This is because of how the glass was made in the days when the manufacturing was more crude, and nothing to do with the 'flow' of the glass.

Panes of glass used to be made by spinning a large lump of glass on the end of a steel rod. As it was spun, it flattened out into a disk which was then cut up into smaller panes. But the bits on the edges were always thicker than those near the middle so the panes often varied in thickness. In fact the very edge section was often so thick you couldn't really see through it and this glass was sold off very cheaply.

When you look in a car **mirror**, why is the view much more **distinct** and the colours **sharper?**

One curious thing about most car side mirrors is they are convex, so not only do things look smaller but objects that are far behind you appear to be a relatively short distance away. If someone is slightly short-sighted, they may not be able to focus their eyes properly on the distant hills, but they will still be able to focus their eyes on the image of the hills in the convex mirror. That is one possible explanation for the effect.

Here's another. We see things by forming an image on our retina, the sensitive coating on the inside of our eyes, and this is made up of two different types of light-sensitive cells called rods and cones. The rods are highly light sensitive, but not very colour sensitive. The cones, on the other hand, are less sensitive to light but are very sensitive to different colours. The rods and cones are not uniformly distributed across the retina. In fact, at the spot near the centre of the eye is a region called the fovea where there are very few rods and the cones are thinner and more densely packed. So this region is more colour sensitive and, because of the higher density of cells, gives a sharper image. It is possible that things viewed in this confined area, as when looking at a reflection in a mirror, will look sharper and the colours brighter. The image of the world just to the side of the mirror will fall outside the fovea and so be less sharp and colourful. Just a theory – I don't think it has ever been proved.

Why are **rainbows** curved?

Rainbows form when light from the sun enters a raindrop and is bounced around inside it. The light takes quite a complicated path; it enters the droplet, bends due to refraction, bounces off the back surface of the droplet and then emerges, having been bent once again as it leaves. For

all this to happen, the light has to hit the raindrop at a certain angle, which is why rainbows don't fill the entire sky.

Why are they curved? Try an experiment. Take a long piece of string and tie one end to a point on the floor about 2m away. Imagine that this end of the string is the sun, and the string itself is one ray of light coming from it.

Now, stand on the other side of the room and hold onto the other end of the string with your right hand so it is taut. Take your left hand and grab a bit of the string about an arms' length away. Keeping the string taut, bring your right hand down so that the string is kinked and points back towards the 'sun'. Lower your hands to the floor until your right hand hits the ground. You should now have something that looks like the path taken by a beam of light from the sun. The bit of string in your left hand is where the light gets bounced by the raindrop and the bit in your right hand, on the floor, is you. The angle made by the two lines of string represents the angle necessary to get a rainbow. If this angle is not constant, you won't get a rainbow. The only way for this angle to remain constant is if you move your left hand (keeping the string taut) to the left and right in an arc. What shape do you get? A semicircle – the shape of a rainbow.

AC TO DC

I know that **electricity** is about electrons **moving**, but what exactly is an **electron?**

An electron is the smallest bit of electrically charged stuff that can be found in everyday matter, and is the tiniest bit of anything you might come across. It is so small that a staple probably has about 5,000 million, million electrons in it. It is much smaller than the atom, of which it is a part, and is the smallest unit of electrical charge.

The electron is so small that the normal rules of sensible everyday things just don't apply. Most people imagine electrons as tiny balls whizzing through a wire, or in orbit in an atom, like miniature satellites. But the picture of electrons as tiny charged billiard balls doesn't always work because they aren't really like that. In fact, early twentieth-century scientists thought of electrons as 'waves of chance', and modern theoreticians in this field think of them as 'fields of multi-dimensional vectors'. And if you want to know what that means, you need a fatter book than this.

Why bother making the **electricity** go **one way**, then the other, in an **alternating** current? What's wrong with it just going **one way?**

You're right – the two main types of electric current are alternating current (AC) and direct current (DC). In DC, the electrons just move along the wire in one direction. But AC has one big advantage over DC – you can change the voltage by transforming it, which makes it more efficient to distribute. Remember, the electricity which arrives in your home may be at 240 volts (in Europe), but it will have been carried down the overhead lines at many thousands of volts. You couldn't do that with DC because you can't change its voltage with a transformer, and the power losses between your house and the power station – which can be hundreds of miles away – would be huge because those losses are related to the square of the current, not the voltage. So, for maximum efficiency, we send power around the country at very high voltages, but low currents.

Also, AC is easier to generate than DC. At its simplest, AC electricity is made by spinning huge coils of wire inside giant magnets. The coil is going in and out of the magnetic field, and this generates a current that goes first one way then the other.

Which is more **dangerous**, AC or **DC?**

Thomas Edison, the great American inventor, believed AC was much more dangerous than DC, which he thought would be the best method of distributing power to homes. To prove his point, he noted that the electric chairs used for execution were supplied with AC. In fact, when it comes to danger it's the quantity of electricity, not whether it's AC or DC, which makes the difference.

However, it is not clear cut. It can be argued that because DC flows in only one direction, this could affect your muscles which are operated by electrical signals. A strong DC signal from outside the body could cause your muscles to contract very strongly causing your hands, for example, to become 'stuck' when holding onto the source of electricity that is giving you the electric shock allowing more and more electricity to affect your body – there would be no escape. With AC that's less likely to happen because the current is always changing direction; your muscles would tighten then relax very rapidly and you would be more likely to be thrown away from the source – in theory.

All electricity, AC or DC, has some dangers, and over about 30 volts there is a risk of electrocution from both types.

When someone gets **electrocuted**, what kills them? Is it the **current** or the **voltage?**

The current. Most deaths associated with electricity happen because the electricity upsets the way the heart beats. Heartbeat is controlled electrically and stray electricity messes up the very delicate balance needed to keep your heart beating steadily. If there has been too much current for too long, the heart can't get its act together again and goes into ventricular fibrillation – a sort of fluttering that is useless at circulating blood. Without blood your brain gets starved of oxygen and you die.

High voltages are more dangerous than low ones because they can drive a high current through your body. However, some sources of high voltages are not dangerous. Devices such as a van der Graaff generator, which produces very high voltages of static electricity, or the spark used to light gas on some cookers, can generate several thousand volts, but there are relatively few electrons involved so only a small current can flow for a very short time.

Why do you have to be **touching** the **ground** to get an electric **shock?**

Electricity is the flow of energy along a wire or through a conductor. If there's no flow, there's no electricity. If you aren't touching the floor, then there's nowhere for the electricity to flow so no electricity will pass through you.

But birds stand on **high voltage** electricity **wires**. Don't their **two feet** make a **circuit?**

Yes birds do make a circuit, but a pretty poor one. As in any circuit, the electricity tries to find the easiest route to travel, and since the bird has a very high resistance the electricity travels straight through the highly conductive wire and hardly any of it directly through the bird.

Although the potential difference between the wire and earth may be thousands of volts, the potential difference between the bird's two feet is extremely tiny and therefore there won't be much of a current flowing between the feet. However, if the bird managed to keep one foot on the wire and placed the other on the ground, you'd pretty soon have a fried bird.

Why do **transformers** hum?

Transformers that are connected to mains electricity hum because the electricity, which is alternating current (see above) is moving to and fro and the 'atomic magnets' inside the soft iron core of the transformer are constantly re-aligning themselves, first one way and then the other, fifty times a second.

Wires carrying electricity have electromagnetic fields surrounding them, and in a wire carrying an *alternating* current these fields will change direction every time the direction of the electricity flow changes. The hum occurs because of these changing fields, and is caused by the vibration of the

transformer box, or covering, because, as the fields change, they attract and repel any metal components in the box or other wires. This movement is the vibration that causes the humming and is normally at 50Hz or cycles per second, which is the same frequency as the alternating current (60Hz in the US).

Why do **power stations** put so much effort into **heating water** up, and then use **cooling towers** to make it **cold** again?

Power stations produce energy in two stages to ensure that as much energy as possible is extracted from the fuel. The first stage burns the oil or gas to produce hot gases which drive the first of two turbines. Once these gases have passed out of this first turbine they are still hot, and this heat energy is used to drive the second process in which these hot gases heat up cold water.

The water turns into steam, which passes through superheating coils to heat it to over 500°C. It then passes into the second set of turbines where it expands, and this expansion drives the second turbine.

Once the steam has expanded it has no useful energy left – not enough to turn a turbine – but it remains hot. So it is cooled to return it to a liquid state so it can be pumped back into the heating cycle and used again. The cooling towers are where the water/steam loses this 'extra' heat.

TANGLED BEDDING TO TASTY TOAST

What makes all your **laundry** get **inside** a **quilt cover?**

Blame statistics. When you stuff a quilt into a washing machine you create lots of folds which act like gates, opening and closing at random as the quilt moves round and round in the machine. Once your laundry finds its way into a quilt it only has two ways to go; it can either move further in or further out, providing a gate, or fold in the quilt, is open. However, there is one gate in a quilt that is always open and that washing can always get into, which is the entrance slit – making it is very easy for washing to get in.

Once in, it then has an equal chance of either coming straight out again or becoming trapped. Once it gets trapped, the chances of it then getting out become less and less as more and more folds would have to be negotiated. If it does get out, the chance that it would get back in again before the end of the washing cycle, are high. So statistics and the laws of probability basically decree that your clothes will always end up inside your quilt cover.

Why is it so difficult to **iron out** a **crease** when it's so easy to **make** a **crease?**

The fibres which make up the materials of our clothes prefer to lie in straight lines, and when you iron in a crease you are aligning the fibres into straight lines. It's not very difficult to do because that is what the fibre would naturally want to do.

Getting rid of a crease is harder because the fibres like being in straight lines, and resist any efforts to disturb them. However, you can make the fibres relax and take them by surprise with the iron – all you have to do is wet them first.

Why is sodium **bicarbonate** used to **clean** fridges and **freezers?**

Most smells from food sources are acidic. Sodium bicarbonate is the opposite – it is mildly alkaline. This means the two can react together to neutralise the acid and get rid of the smell. For example, butyric acid which might come from some gone-off food, will react with sodium bicarbonate to make sodium butyrate, which doesn't smell. However, some smells, like those from fish and meat, are alkaline so sodium bicarbonate won't get rid of fish smells. If you want to get rid of these from your kitchen, some say you should leave an open bowl of vinegar (acidic) there overnight.

Why does **newspaper** go **yellow** faster than other **paper?**

The newspaper publisher's main concern is to get the news to the reader as quickly and as cheaply as possible, so they use the cheapest papers and the cheapest inks. Newspapers have a very short shelf life – there isn't much point in making them out of posh paper because they are likely to be thrown away the day they are bought. The main components of wood are cellulose and lignin. Cellulose is long-fibred and strong, which is why paper remains supple over a long time. Lignin (or sap) is what makes wood hard, and is acidic. In expensive papers, the pulp is cooked to remove the lignin, but newspapers are made from wood that hasn't been cooked so most of the lignin remains, and it is this which turns yellow on exposure to sunlight.

Why do paper towels **absorb water better** than other types of **paper?**

Paper is made from tree fibres that are naturally hydrophilic, which means having a strong affinity for water. In fact, they love it. But the last thing you want are those tree fibres sucking your pen dry when you're using writing paper, so clay is added to the paper-making mixture to stop that happening.

Paper towels don't have these additives so are essentially basic paper fibres with only a few additives to improve strength. In addition, absorbent paper goes through a process called crêping. This is where the fibres are fluffed up by being stuck to a drum and passed over a knife edge. This pulls fibres away from the surface and buckles the paper. Crêpe paper is made this way and is the most extreme form of crêping. Paper towels go through the same process but not to such an extent.

Why do **vegetables** go **soft** when you **cook** them?

All plants are made up of millions of plant cells, and each plant cell is surrounded by a very strong cell wall. Adjacent cells are held together by a glue. When you eat a raw vegetable or an unripe piece of fruit, your teeth need to break the glued cells apart and this is what gives the crunchy feeling. Cooking softens the glue between the cells so that your teeth slide easily between the cells. The same thing happens when fruit and vegetables turn ripe.

Why can't you cook **rhubarb** in **aluminium pans?**

Aluminium forms a thin oxide layer on its surface to protect the metal from destruction by the atmosphere, but certain foods, because of their high acidic or alkaline content, easily penetrate to the metal's surface and form aluminium compounds which may be grey or black in colour. Rhubarb is highly acidic with a pH of 3.1; its leaves have very high concentrations of oxalic acid, and for that reason are considered poisonous and shouldn't be eaten. So one good reason not to cook rhubarb in an aluminium pan is because you'll spoil the pan, and some of those dark compounds may discolour your food.

The other reason is that some of the aluminium will find its way into the food, although not in huge quantities. No research has ever suggested there is any harm in using aluminium cooking pans. Anyway, there are many other sources of aluminium in our diet that cooking pans are possibly the least of it. Tea, herbs and spices have particularly high concentrations of aluminium, but much of it passes straight through the body, unabsorbed.

Are **green** potatoes **poisonous?**

The green colour you see when potatoes have been left out in the sun is chlorophyll, which in itself isn't poisonous. But potatoes contain a chemical called solanine which *is* poisonous, and this becomes concentrated just under the skin when the potato has been left out in sunlight. It is most concentrated in the eyes, skin and sprouts, which is why we often peel potatoes to remove them. Solanine is also found in poison ivy and in aubergines. It can cause stomach upsets, but in small quantities it will not kill you. You would have to eat a large quantity of green potatoes before you felt ill. An average person would have to take in 2lbs (just under a kilo) of fully green potatoes before suffering any poisoning effects.

What is it that makes the **fizzing** sound when you put **food** into a **hot pan?** Is it a chemical **reaction?**

This sound is simply the sound of small drops of fat and water spitting at each other. As you heat something up, the fat melts and drips down. The water also escapes, but as the two can't mix, and since both are rather hot, they end up trying to get out of each other's way rather violently. This rapid movement is the noise that you can hear.

What makes a **substance** have a **smell?**

Things smell because they release small molecules into the air. These molecules enter our noses where little receptors capture the smells and recognise them. These smell receptors send a signal along nerves to the brain and the brain identifies the smell by recognising which type of receptor the nerve signal has come from. Because receptors are certain shapes, only certain shaped smell molecules can fit in these holes so only the right types of smell can enter the right receptors. Some objects give off smells but our noses don't have the right shapes for them to fit into so we can't smell them, but some animals can.

What's the **smell** you get from **freshly** rained-on **tarmac?**

These smells are most obvious when rain or a hosepipe wets a hot surface so that there is a very strong burst of evaporation of surface water into warm, upward-moving water vapour. This burst carries with it traces of gases and vapours from the particular surface. Roads for example, have accumulations of oil and rubber, while earth and lawns have many sources of gases. So what you smell is this sudden,

126

highly enhanced evaporation carrying an assortment of surface material. The smell will be stronger after a dry spell because more surface material will have collected, and even stronger in hotter weather when the burst will be more pronounced.

You don't only get that 'just rained' smell from tarmac. It can happen in cities, the countryside, and certainly the African bush. The smells that you get from different ground surfaces are all going to have similarities. The only difference will be a change in the exact composition of the volatile gasses that make up the smell.

People also talk about a 'smell before a thunderstorm', which is all to do with electricity in the atmosphere. Thunderstorms carry a lot of electrical charge and the moving air quickly builds up static electricity. It is this electricity some people say they can smell.

Why does **salt** make you **thirsty?**

The body is made up of about 60 per cent water, some of which is contained in our cells, but part of it is being used to carry the blood cells around your body. When you take in a lot of salt with your food, it finds its way into this 'extra-cellular' water in your body. You are now in a position where the main cells of your body contain low salt water, while the water outside the cells has become much saltier. So, water flows from inside the body cells to join the sodium in your bloodstream. Your cells are becoming drier, you feel dry, so you drink. The more salt you've taken in, the more you will have to drink to quench your thirst.

Why do boats **float?**

It's time to remember the great discoverer, Archimedes, whose famous principle states that when a body floats, the upward force of buoyancy is equal to the weight of water it displaces. He came to this conclusion in his bath and, it is said, was so excited at this fundamental scientific breakthrough, that he leapt out of the water and ran, unclothed, through the streets shouting 'Eureka!'

A steel boat is very heavy, and you might well expect it to sink; certainly, if it were just a lump of steel of the same weight it would sink to the bottom of the sea faster than Archimedes could shout 'Eureka'. But suppose the boat weighs 1,000 tons and is lowered gently into the water – once it has displaced 1,000 tons of water, it will then float *providing it is not submerged.*

To ensure it does not sink, the boat has to be the right shape. If you were to put a plank of wood on the water, it would float because all the trapped air between the fibres of the wood mean that it is less dense than the water. Because of the shape of a ship, it also contains a lot of air. So, if you work out the average density of the volume of air contained within the hull and the steel combined, you might end up with something of surprisingly low density which floats. Of course, once the water leaks into the ship the average density quickly changes and, if it becomes greater than the water around it, the ship will sink.

Is it true that when you pull the **plug** out of your **bath**, the water always **swirls** down the plughole in a **clockwise** direction?

No, it won't. It will depend on how the water is moving while still in the bath. The belief that water goes different ways down a plughole depending where you are on the surface of the Earth is a myth. The Coriolis force, which results from the rotation of the Earth and decides which way storms rotate (anti-clockwise in the northern hemisphere, clockwise in the southern) only works on oceans or atmospheres. It has no interest in your bath water. More likely to determine which direction your bath empties is the shape of the plughole and what the water was doing the instant you pulled out the plug.

How can soft **leather** sharpen a **razor?**

As a razor blade is used, its edge is dulled by a 'burr' which forms on either side. This effectively widens the blade and makes it blunt. By dragging the blade across a leather strop, the soft burr can be removed to recreate a thin edge, and like magic your blade is sharp again. Strops aren't so effective on thicker blades, such as scissors, because the bulk strength of the material is greater. However, strops can be very effective on carbon steel knives because carbon steel is weaker than stainless steel.

Will a hot **squash ball** bounce higher than a **cool** one?

Certainly a ball which is more flexible will bounce higher because, as the ball hits the floor, kinetic energy is stored in the ball as it is 'squashed' on the side hitting the floor. This energy can then be released as the ball expands again, sending the ball back up into the air. The more kinetic energy is stored in the downward movement, the more is released as it bounces and the higher it goes. Think of a spring. If you compress it and then let go, the energy you put into compressing it will be released as it springs out again – as the ball hits the ground the material of the ball is compressed against the floor and then released suddenly, flinging the ball back up again.

So why does the rubber of the squash ball behave more like a spring the more you heat it up? Rubber is called an elastomer, which means it consists of molecules which are coiled up, a bit like springs, and can be stretched and squashed just like a spring. The more you heat up the rubber, the more flexible the bonds between the molecules become and the more bounce you get.

Does the **pressure** in a **squash ball** affect the rate at which it **bounces?**

During a game of squash, the kinetic energy of the squash ball is converted into heat energy as the ball collides with the walls and the racket. This causes the temperature of the air inside the squash ball to rise, and so the air pressure inside the ball will increase. For something to bounce it must be made of a material which changes shape on impact and then tries to returns to its original shape. An increase in the air pressure inside the squash ball makes the ball return to its original shape faster, and therefore makes it bounce higher.

Why is it harder to **pedal** a **bike** with **flat tyres?**

It's all to do with friction. When a tyre is fully inflated, only a small portion of it is in contact with the ground, causing less friction between the tyre and the ground than when the tyre isn't inflated. This is also the reason why a road bike (or racer) is easier to cycle on the road than a mountain bike. Mountain bikes have thick wheels with a lot of the tyre in contact with the road so they can grip well when cycling through mud. Road bikes don't need tyres with such good grip, so the amount of the tyre in direct contact with the road is much smaller – sometimes as little as a few millimetres.

Why does **polish shine** when **brushed?**

A shine is merely a good reflection of the light falling on a surface, and so you can only get a good shine from a very smooth surface. If the reflected light rays are all parallel to each other, you get a sharp reflection. Mirrors are an example of a very smooth, highly polished surface. A scuffed surface, on the other hand, will reflect the light in lots of different directions and so you won't get a good reflection.

Most polishes are based on waxes, which are long chained hydrocarbons, and are very good at getting into the many small nooks and crannies that exist on surfaces. When you apply polish to a surface, you are filling up all of the imperfections with layers of wax. Then, by brushing the polish, you are creating a smooth surface which can reflect the light brilliantly making it appear to shine.

Why does **toast** taste **different** to **bread?**

When you toast bread, two things happen. Firstly, as a result of the heat the water in the bread evaporates which dries out the bread. Secondly, sugars on the surface of the bread begin to caramelise and the sugars and proteins on the surface begin to react with each other. These reactions result in the bread turning brown. Caramelisation is the name given to the extensive chemical reactions that occur when any sugar is heated to the point when it begins to break apart. More than a hundred different compounds, many of which have distinct flavours, are made during caramelisation, and it is these compounds which gives toasted bread its new flavour.

4

HOW

on
Earth . . . ?

Spinning Planets to Atomic Bombs

Lightning Bolts to Volcanoes

Thirsty Oaks to Massive Mushrooms

SPINNING PLANETS TO ATOMIC BOMBS

How does the **Earth** move?

The Earth moves because of the gravitational pull between it and the sun. Isaac Newton, in 1687, first explained that any two objects will attract each other, and the force of attraction was related to the mass of each object and their distance apart. It is this force of attraction between the planets and the sun which keeps them in orbit, and without it the planets would be able to fly off in a straight line, just like a conker when you let go of the string.

In fact, the way the Earth goes round the sun is very similar to the way a conker can be swung round on a piece of string. Without the string the conker would just whiz off. So why doesn't the Earth eventually fall into the sun? Because the Earth is also moving forwards, so while gravity is pulling it towards the sun its forward movement is trying to make it escape. The result of these two forces causes the Earth to orbit the sun.

I have heard that the **Earth** is 5 **billion** years old. How can we possibly **know?**

In the nineteenth century scientists attempted to work out the age of the Earth by estimating how long it would take molten rock to cool down to its present temperature, and on that basis Lord Kelvin decided that the planet was no more than 100 million years old. This didn't tie in with

Charles Darwin's new theory of evolution, and the discrepancy wasn't resolved until the discovery of radioactivity when it became clear that the minerals in some rocks were being kept warm due to radioactive decay. The rate of the radioactive decay became the clue to measuring the age of the Earth.

The rate at which radioactive isotopes decay is often expressed as their half-life, which is the time it takes for one half of the isotope to decay. In some elements this can be rapid, but an isotope commonly found in rocks is called uranium-238, and this has a half-life of 4.5 billion years. So in theory, if you can measure how far these isotopes have decayed, you can calculate how old they are. The process, however, is much more complex since the uranium goes through several other radioactive stages, all with much shorter half-lives. Uranium-238 ends up as lead, but uranium-235, with a half-life of 0.71 billion years, ends up as lead-207. By measuring the concentrations of these two isotopes of lead in rocks, geologists can measure their age.

For measuring the age of something more recent, say within the last 50,000 years, the radioactive isotope carbon-14 is used because this has a half-life of only 5,730 years. Also, carbon-14 is present in all living plants and animals until they die, when it starts to decay. This makes it useful for telling the age of trees and fabric, and human and animal remains.

If you were to **drill** a hole to the **centre** of the **Earth** and out the other side, could you **fall through** it?

You would fall 3,965 miles towards the centre, and then your momentum would take you part of the way outwards the other side. But gravity would soon pull you back towards the centre again, and for a few swings you'd oscillate about the core, eventually coming to rest at the centre under no gravitational pull at all. However, you would be fried and nicely crisp by now, given the temperatures at the centre of the Earth are believed to be 3,000°C. And don't forget the pressure, estimated to be 14.2 million times atmospheric pressure.

How **big** is the **Earth** and how can we **measure** it?

The facts are these: the Earth is 7,926 miles in diameter, or just over 24,000 miles in circumference, around the equator. We've known roughly how big the Earth is for quite some time now. It was 2,000 years ago that the Greeks first worked it out.

A man called Erastothenes said, 'The Earth is round, right?' and everyone said, 'Yes, so?' Erastothenes replied, 'Well if I know the Earth is round, and if I assume the Sun is a long way away, the light rays from the Sun are parallel by the time they reach the Earth. So using the geometry of triangles that Pythagoras has just come up with, I can work out how big the Earth is.'

He knew that at midday in Aswan, where he lived, the sun was directly overhead because the sun's rays reached to the bottom of the town's well, which can only happen if the sun is right overhead at the time. He also knew that a town called Alexandria was directly north from Aswan. What he needed to do was find out how far away Alexandria was, and what angle the sun cast a shadow at midday.

So, he got an army to march between Aswan and Alexandria and measured the time it took so he could work out the distance – it was about 500 miles. This distance was only part of the total distance around the Earth, of course. But if he could work out what *fraction* of the circumference of the Earth this distance was, he could work out the total distance around the Earth. But how did he work out that crucial fraction?

This is where all the angles of shadows comes in. The Earth is curved, so if the light from the sun is hitting the Earth at right angles at Alexandria then it won't be at right angles anywhere else. Try this: draw a circle with an arrow pointing straight at it. Next to that arrow draw another, and another, until you have lots of parallel lines pointing at the Earth. You'll notice that the angle that the lines make when they touch the circle changes. It is this angle that Erastothenes was trying to measure at Alexandria. Using Pythagoras's laws about the angles of triangles, he could work this out from the height of the stick and the length of shadow cast.

Draw another circle, and then a line from the edge of the circle to the centre. Then draw another line from somewhere else on the circle to the centre. You'll now have two lines touching at the centre of the circle. These lines have an angle between them and it is this angle that Erastothenes was measuring. Erastothenes knew the distance between the two points from how far the army had marched, and he knew the angle at the centre of the circle from the shadows. From this he could work out the size of the Earth because he knew that there were 360° in a circle. He actually got very close to the results given by modern methods of measurement using satellites, which is amazing when you remember this all happened 2,000 years ago.

Is it true that
the **Earth's** population
could **stand** on
the Isle of **Wight?**

It was true once, but not now. The Isle of Wight is approximately 400km². If you allow 1,000cm² per person, which is pretty tight, then you can fit about 4 billion people on the Isle.

In 2006, the world population reached 6.5 billion which would have overcrowded the Isle of Wight, but you could still put the entire population of the world on the Isle of Wight and the Isle of Man combined.

What was the total number of
dead people in the **past**
compared to the number of **people**
alive now?

Estimating the number of people that have lived since the world began is extremely difficult. Not until the middle of the seventeenth century is there any record of an estimate, which was made by a Dutchman, Isaac Vossius, a scholar and manuscript collector, who said the population of the world was 545 million people. Since then the population of the world has generally increased, despite setbacks caused by wars, epidemics and famines.

A major change took place in the seventeenth century with a decline in the death rate as a result of improved hygiene practices, better food supplies, increased vaccinations and the disappearance of diseases such as the plague and cholera. Together with a steady birth rate, the population of Europe grew at an enormous rate.

The difficulty with working out the population of the world since human life began is that it is impossible to tell how many people lived in prehistoric times. If we assume that human beings (difficult to define at the early stages of human development) have been around for the last 200,000 years, and the population increased linearly until the year 8,000BC, anthropologists have estimated that prior to 8,000BC, 768 billion people have lived on Earth, and since AD8,000, 2,207.5 billion. This adds up to 2,975.5 billion people who have lived since human life began – but remember there is a lot of guesswork involved in coming up with these figures.

The current population of the world is 6.5 billion.

If the **centre** of the **Earth** is so **hot**, why isn't the **ocean warmer?**

The centre of the Earth is extremely hot – about 4,300°C. But that temperature falls the nearer to the surface you

get. In fact, the crust of the Earth – the bit round the outside – can be quite cool if it isn't near any volcanic areas. So while the ocean is above the crust, and you'd think that the heat from the Earth would escape into it, it doesn't because the crust is much cooler than the centre of the Earth.

What **holds** the **sky up?**

This question isn't as silly as it sounds! The atmosphere is a gas, and gases can be squashed, so why doesn't gravity pull all the air down to the ground so that the atmosphere is only a few metres thick?

Air stays where it is because of the movement of its molecules. In solids they don't move very much, in liquids they move a bit more, and in gases they move around a lot. All of this movement fights the gravity which is trying to pull the gas to the ground. The gravity is strong enough to stop the air escaping into space, but isn't strong enough to battle against the movement of the molecules.

How **high** does the **atmosphere** go?

The atmosphere extends up to 1,500 miles or 2,414km above the surface of the Earth. However, the bulk of the gases (about 75 per cent) occur within the first 10 miles (16km) above the surface.

The atmosphere is divided into layers. The exosphere occurs above 435 miles (700km) and is really at the fringes of the atmosphere. Then the thermosphere between 55 and 435 miles (85–700km), the mesosphere between 30 and 55 miles (48–88km), and the stratosphere between 7 and 30 miles (11–48km). The stratosphere contains the ozone layer which protects living organisms on the Earth from the harmful ultraviolet rays of the sun.

The troposphere averages about 7 miles thick, but only 5 miles at the Poles and 10 miles at the equator. This is where the weather takes place.

What **effect** would a **nuclear bomb** have on the **weather?**

Nuclear bombs suck a lot of dust and rocky debris into the atmosphere. Most of this rock and dust falls back to the Earth, but some of the finer dust can stay in the atmosphere for weeks, or even months. One result of this is red sunsets as the dust scatters light from the sun, but a much more subtle effect is that of global cooling.

The dust high up in the atmosphere reflects a lot of the light from the sun so that it never reaches the Earth's surface. This causes the Earth to cool down. This can also happen when volcanoes throw dust into the atmosphere. In fact, it is thought that the sudden fall in the rate of global warming at the end of the 1980s was due to the eruption of Mount Pinatubo in the Philippines.

Why are **hydrogen bombs** so much more **powerful** than **atom bombs?**

An atom bomb is a fission bomb and works by splitting the atomic nucleus. A hydrogen bomb is a fusion bomb and works by combining atomic nuclei.

A hydrogen bomb contains a ball of plutonium-239 about the size of a grapefruit. An explosive chain reaction can start in this sphere whenever a stray neutron enters a plutonium nucleus, but it rapidly fizzles out because too few of the neutrons find other plutonium nuclei. To make sure the neutrons make sufficient direct hits, the ball of plutonium-239 is surrounded by charges which, when exploded, compress it into a much smaller volume. The uranium nuclei are now closer together, the neutrons have a greater chance of collision, and the chain reaction starts. (A neutron source ensures there is a neutron there to set off the reaction.) The result is a terrific burst of heat energy equivalent to about 20,000 tonnes of TNT.

In a hydrogen bomb, two heavy hydrogen nuclei fuse together to produce a helium nucleus, and this reaction is triggered by a fission bomb which is placed in a jacket containing heavy hydrogen (deuterium). The resulting bang is equivalent to at least a million tonnes of TNT.

Fission energy is released in a controlled way in a nuclear reactor but nobody has yet managed to generate power by controlled fusion. The sun and other stars, however, get their energy from nuclear fusion.

LIGHTNING BOLTS TO VOLCANOES

How many **slices** of **bread** could you toast in a **lightning strike?**

One strike of lightning has about 10,000,000,000 Joules of energy (10 billion Joules) and temperatures within it can rise to about 30,000°C, and last about 10 milliseconds – 10 thousandths of a second.

A typical toaster works at 450°C for two minutes to toast two slices of bread using 25kW of energy. One lightning strike is 100 million watts, or 100,000kW, so each lightning strike has the power of about 80,000 toasters and could toast 160,000 pieces of bread. Of course, lining up 160,000 pieces of bread and turning them over just at the right moment could be tricky within only 10 milliseconds to play with.

Does a lightning **strike** start in the **sky** or the **ground?**

A thundercloud contains billions of water droplets and ice crystals. Scientists think (although they are not sure) that static electricity builds up inside a cloud when these droplets and crystals crash together. As they do so they exchange some of their electric charge. The larger drops gain a negative charge and fall to the bottom of the cloud so you end up with a strong positive charge at the top of the cloud, and a negative charge at the bottom. The air, a good insulator, keeps them apart until the charges become too great, which is when the insulation breaks down and a flash of lightning jumps between the two. This is by far the most common kind of lightning. Ten per cent of all flashes occur within clouds.

But those descending negatively charged droplets have an effect on the ground as well, by repelling negative charges on the buildings, trees and land below leaving a positive charge at ground level. A negative electric potential now exists between the cloud and the Earth. If a big enough charge builds up then the insulating layer of air can break down again. When this happens the charge at the base of the cloud discharges itself by seeking a path to the ground with flashes, called leader strokes. As the leader stroke gets close to the ground, a large positive charge, called a streamer stroke, builds on the ground which then rises until it meets the leader stroke about 10–20m above ground level. These strokes then create a channel along which a second more powerful flash can run – a return stroke. So a lightning strike can start in the sky *and* on the ground.

Why doesn't **lightning** affect **people** in a **car?**

The car works like a Faraday Cage. Michael Faraday demonstrated, in the early nineteenth century, that an electrostatic charge only appears on the exterior surface of a charged conductor, and not on the inside. In 1836, he built a room within a room, the smaller one coated with foil. He then sent very large electrostatic voltages across

the outer room and measured the electrical charge on the inside of the foil-clad room. He found none, and that has become known as the Faraday Cage effect. It is still used to protect sensitive electronic equipment, and is one of the reasons your mobile phone would not work if you took it into a foil-clad room – not only would there be no electrostatic charges, but electromagnetic waves wouldn't pass either.

Not only cars act as Faraday Cages; aeroplanes make good cages too which is why planes are frequently hit by lightning and the passengers don't feel a thing. And since the same principle applies to contained radiation – it cannot escape from an enclosed conductor – it explains why radiation does not pass through the metal walls of a microwave oven.

What generates the **noise** of **thunder?**

A lightning flash heats the air around it to about 1,000°C, about a sixth as hot as the surface of the sun. As things get hot they expand and this rapid heating of the air causes it to expand quickly causing a shock wave. This is the thunder.

The reason we hear different types of thunder, such as long rolls or short cracks, is due to the path the lightning takes. Imagine a flash which starts 1 km above our heads and travels diagonally to hit the Earth 1km from our feet. Every point on that path is (very roughly) the same distance from us, so the sound from every point of the flash hits us at about the same time which gives a loud 'crack'. Now, imagine a flash that starts about 1km up and ends close to our feet. The sound from the start of the flash's path has to travel much further than the sound from the end and will, as a result, reach us much later. This creates the long, slow roll of thunder.

Can you get **lightning** with **no thunder?**

Thunder is caused by lightning: the lightning heats up the air around it to great temperatures. When this happens the air expands, and the molecules which were close to the lightning are forced outwards, banging into those next to them. These in turn bang into the next lot of molecules and this process continues until the sound reaches your ear.

If you could create lightning where there was no air or anything else to carry the sound (perhaps in outer space) then you would not hear thunder. However, any lightning occurring in the Earth's atmosphere would have to travel through air, thereby heating it up and causing a sound to be produced.

153

The other factor to consider is that the sound of the thunder dies away with distance because the air molecules gradually lose energy. So if you were in an area of very flat terrain you might be able to see lightning, but the thunder would no longer be audible to the human ear by the time it reached you. But the thunder would still have been produced, it's just that you wouldn't be able to hear it.

Why does it get **colder** as you go up a **mountain** even though you're getting **closer** to the **sun?**

Firstly, you're only getting a little bit closer to the sun, and since the sun is 93 million miles away and the tallest mountain on the Earth, Everest, is only about 5½ miles high, you are not getting much closer to the sun by climbing all the way to the top. So forget about closeness to the sun.

Instead, think of a bicycle pump. When you pump hard, the pump starts to feel warm as you compress it. Then think of a fire extinguisher which has just gone off: the pressure inside it has been rapidly reduced and it feels cold, sometimes cold enough to appear frosty. Expanding air cools, compressed air heats up.

Also, hot air rises. When the sun strikes the surface of the Earth and warms it, the adjacent air is heated and starts to

rise. But as it rises it experiences a change in atmospheric pressure which is about 15lb per square inch at sea level, and decreases with altitude (because there's less atmosphere to press down on us). By the time the warmer air has risen to 18,000ft – mountain top height – the atmospheric pressure has roughly halved, allowing the air to expand. Expanding air cools.

Other factors can help to make mountain tops colder. Snow, ice, and to a certain extent bare rock, reflect much more heat from the sun than forests and fields found at lower altitudes. So forests absorb heat, collecting it and making the area around the forest warmer, while the snow and ice fields can reflect the heat back towards the sun.

And there's one other reason mountain tops are colder: clouds act as a very efficient blanket, especially at night, and help keep the atmosphere warm by reflecting back to Earth any heat that might try and escape. High mountain tops tend to be above the level at which clouds form and so don't benefit from that night-time protection.

If the **ice** around the **North Pole** floats, why doesn't it **move around** with the **winds** and **tides?**

It does. The ice drifts from Russia, across the Arctic Basin, and towards Canada. If you are going to walk to the pole, there are big advantages to leaving from the Russian side because the drift will be with you and you'll get there sooner.

The vast majority of the ice-covered part of the Arctic Ocean is almost landlocked by Russia, Greenland, Canada and Alaska and the openings to the south are blocked by underwater mountain ranges. So the sea ice is somewhat limited in where it can go and ends up circulating around its own basin.

156

If it's **12 noon** here, what **time** is it at the **South Pole?**

The simple answer is that the Poles do not fall into any one timezone. International timezones are designated as being along the line of longitude, but these all meet at the North and South Poles, so you couldn't say you were in any single one of them. In theory, the lines of longitude not only converge at the Poles, but they also diverge away from them. This means that, in theory, the time will change depending on the direction in which you look.

This doesn't help anybody, so the people who live and work at the Poles agree that the time there is the same time as GMT. So if it is 12 noon GMT in Britain, it is also 12 noon GMT at the South Pole.

If you tried to put up a **signpost** at the **North Pole,** how would you know where to **position it** so that it **points** towards the east?

Trick question? I think so. From the North Pole, whichever way you point you are looking south.

Does a **volcano** have any **good uses?**

Yes, there are lots of good things to be said for volcanoes. In volcanic areas, water contained within the rocks can be heated, and that water used to generate hydrothermal energy. In Iceland, they use it in greenhouses to grow tomatoes and even bananas – yes, Iceland is Europe's largest banana grower! Hydrothermal energy can also be used to generate steam to power turbines and produce electricity.

Soils produced from the ash by the weathering of debris from volcanic eruptions are extremely fertile and often used for agriculture.

On the island of Lanzarote, to the west of Africa, the locals use volcanic rocks as building stone; it's readily available, cheap, and there are very few other building materials. It is also light, so easy to transport, but not very strong, so buildings can't be very tall – which is great if you don't want to ruin your skyline with skyscrapers!

Volcanic eruptions themselves have been important in geological history by introducing gases into the Earth's atmosphere, in particular water vapour, sulphur, carbon dioxide and carbon monoxide. These gases have played an important role in the development of our atmosphere and therefore the very existence of life on Earth.

Finally, volcanic activity produces fluids in rocks which are rich in minerals. When these fluids cool, the minerals crystallise out and are deposited. Many of these minerals are of considerable economic wealth.

Where does **volcanic gas** come from?

The most common volcanic gases are carbon dioxide, water, sulphur dioxide and hydrogen sulphide. Small quantities of other volatile elements and compounds are also present such as hydrogen, helium, nitrogen, hydrogen chloride, hydrogen fluoride and mercury. These gases come from the magma – the molten rock beneath the Earth's surface – and exactly which gases are released depends on temperature, pressure and which other volatile elements are present.

As magma ascends from deep inside the Earth, the pressure acting on it decreases and allows the various gases to 'bubble' out. Carbon dioxide begins to separate at depths of about 40km, whereas most of the sulphur gases and water are not released until the magma has almost reached the surface. It is the expansion of these emerging gas bubbles which tears the magma into clots and forms the lumpy material which volcanoes spit out.

THIRSTY OAKS TO MASSIVE MUSHROOMS

How much **water** does an **oak tree** take up in a **year?**

There are so many 'it depends', such as the condition of the tree, fluctuations in temperature, availability of water, light and air movement. They all make a huge difference. But for an average deciduous tree, about 50,000 litres a year are used.

Why do **flower** petals **open?**

Plants can't move around and make friends so they have to rely on other methods to reproduce, and so they've developed a mechanism called pollination. Pollen consists of the male sex cells of the plant and is simply tiny grains formed by the stamens of a flower. These can be carried in the wind, or more usually by insects, to the ovaries of another plant, which are the female parts of the flower. When this happens, the plant is fertilised.

Petals only open when a flower is ready to be pollinated and the petals attract suitable pollinators, which is why a lot of plants have brightly coloured petals. Yellow, for example, is very attractive to insects, especially bumble bees. However, it costs the plant a lot of energy to make these petals and so they will only open for a short time.

There are certain environmental triggers, such as temperature, which the plant will recognise as being the right time for insects to be pollen hunting, and so the petals will open. No one's really sure quite how this happens, but it seems that changes in the water pressure in the plant are usually involved in the actual opening process.

Do **trees** get **cancer?**

The answer is no, although sometimes it might look as though they do. In areas where they have been damaged, or where a branch has fallen off, they can develop a kind of growth which is in some ways akin to cancer. However, the similarity ends there because once the wound is sealed, the activity of the callus stops, leaving a woody lump on the tree which does it no harm at all. In cancer, the cells just go on proliferating causing organs to malfunction. This doesn't happen in trees.

Why does the **sun** turn **red** as it **sets?**

The sun's changing colours at sunset (and sunrise) are the result of us seeing the rising and setting of the sun through our atmosphere. When the sun is higher in the sky, we are looking almost directly at it, so only a thin layer of air lies in the way. At sunset, though, we are looking towards the horizon through a much thicker layer of atmosphere.

As light from the sun enters the Earth's atmosphere it is affected by air molecules. Individual photons of light collide with these molecules and are bounced off, scattering in different directions depending on their wavelength and colour – in just the same way as water droplets in a rainbow scatter the light into a spectrum of colours. Blue light has the shortest wavelength and is the most prone to scattering, which explains why the sky is blue. At the same time, the removal of the blue light from the sun shifts its colour slightly towards the red, making it appear more yellow than it would from space.

At sunset, the scattering effect increases as the light has to travel a greater distance through the atmosphere. First green, then yellow light are affected, until finally the only light we see directly from the sun is orangey-red.

Another interesting effect of the thickening of the atmosphere is called refraction. The atmosphere acts as a lens, bending the light from the sun's image itself, so that the sun's image lingers above the horizon for a few moments after the sun has actually set.

What is the **fastest** thing on **Earth?**

It all depends on what you mean by thing. Here's a list of some very fast things you can find on this planet, take your pick:

LIGHT. Nothing can ever travel faster than light, so the very fastest thing on Earth is light itself and it travels at 186,000 miles per second. The theory of relativity says we could never travel at the speed of light because it would require an infinite amount of energy to accelerate to that speed.

A SPACECRAFT. Both the Russian Space Station MIR and the American Space Shuttle orbit the Earth at about 25,000mph. I suppose they can't be called the fastest thing on Earth, because they're in space. But over a period of time, satellites and space probes are the fastest man-made objects.

BIRDS. It is very difficult to measure how fast a bird is going since you need to time how long it takes to cover a particular distance. The trouble is that birds are much more interested in swooping all over the sky rather than flying in nice straight lines so we can time them easily. But the spine-tailed swift has been measured by some Russian scientists as flying at 106mph. In Britain our fastest bird is the Peregrine Falcon, whose fastest measured speed in a dive is 82mph. The scientists measured this by attaching little speed-measuring devices to the falcon's wings.

ANIMALS. Just like birds, it's very difficult to accurately measure how fast they go. The cheetah is generally thought to be the fastest animal on land and the best measurements give its speed as 51mph, though many scientists think it can go as fast as 60mph. In an effort to find out once and for all how fast cheetahs could go, scientists put a group of them on a greyhound race track (without the greyhounds) and tried to time how long it took them to complete one lap of the track. Unfortunately the cheetahs just sat down and nobody could persuade them to move. Sometimes being a scientist isn't easy.

What is the **largest** living **thing?**

A fungus called the honey mushroom (*Armillaria ostoyae*) is the biggest organism in the world. It was found beneath the soil of the Malheur National Forest in eastern Oregon and now covers 2,200 acres. It started from a single spore too small to see without a microscope and has been slowly

spreading through the forest (killing trees as it goes) for 2,400 years. No one has estimated its weight.

The blue whale (*Balaenoptera musculus*) is the largest animal in the world measuring about 26m (85ft), but lengths of 21m (70ft) are more common. A blue whale's heart alone may weigh 908kg (2,000lb) – as much as a small car.

The largest plant is the giant sequoia tree; the largest of its kind is the 'General Sherman' at 274.9ft tall and 102.6ft in circumference.

5

SKY

High and Beyond

Twinkling Stars to the Man in the Moon

Crashing Comets to Growth in Space

TWINKLING STARS TO THE MAN IN THE MOON

Does the **twinkling** of the **stars** change throughout the **night?**

Yes it does. Stars twinkle because air currents in the atmosphere affect the light from the stars as it passes through. Just as ripples on a pond produce wavy reflections, moving air makes light appear to flicker. When stars are rising, they are lower on the horizon and their light has to pass through more of the atmosphere, so a star will probably twinkle more than when it appears overhead. The colour of the stars might also be slightly different, due to refraction, but in practice the light is so tiny that with the naked eye you won't notice any difference.

What are **shooting stars?**

Shooting stars are tiny fragments of space rubbish which burn up when they hit the Earth's upper atmosphere, leaving only a short, bright trail of light behind them. Shooting stars can be predicted. They appear regularly as showers once a year as the Earth crosses the trail of dust and debris left behind by a comet. Some can be seen at any time of the year, and these are caused by small particles of leftover material from the birth of the solar system itself. Large shooting stars which survive the atmosphere and hit the Earth are called meteorites.

A fascinating group of meteorites are those which come from other planets. Since the 1980s, hundreds of fragments of rock have been recovered on the ice planes of the Antarctic – a region where rocks could only exist if they had fallen from the sky. Analysis of those rocks has shown similarities with those on the Moon and Mars, so scientists think they must have been flung through space by huge explosions in the distant past and rained onto Earth.

How did the **constellations** get their **names?**

Constellations are mainly used by astronomers these days as a convenient way of finding their way around the sky. They are a hangover from the days when astronomy and astrology were combined, and people believed that the patterns of the stars, and the movements of the planets through them, affected events on Earth. Today, of course, we know that the patterns of stars we see are simply caused by line-of-sight effects between unconnected stars, and that the very slow movement of the stars will eventually change the patterns out of all recognition.

Every ancient culture on Earth seems to have seen fantastic creatures and mythological figures in the stars. The constellations' names we are familiar with today come from a list drawn up by the astronomer Ptolemy of Alexandria in about AD150. His list contained 48 constellations, mostly animals or characters from Greek myths. Some of these constellations fitted together to form entire scenes from the legends – the clearest of these patterns, which dominates the winter sky, shows Orion (the Hunter) battling Taurus (the Bull), while Canis Major and Minor (his hunting dogs) wait behind him.

Ptolemy's system of constellations remained unchanged until the 1600s, when new constellations were added. Then the invention of the telescope meant that it became important to include the fainter stars. This led to a chaotic situation with astronomers fighting to name new constellations. It was finally sorted in the early twentieth century when the International Astronomical Union finally established the set of 88 constellations we know today.

Why don't the **constellations** change if the **stars** are moving so **fast?**

Although the stars are indeed moving very fast at many kilometres every second, they can also be trillions of kilometres away. So they have to travel for thousands of years before we can see any difference in their positions. In one short human lifespan you won't notice a difference in the way the constellations appear, but if you could come back in a few thousand years the sky would look very different. To give you an idea of the apparent motion of the most rapidly moving stars, it would take about 200 years for one of them to move the distance equivalent to the diameter of the full Moon.

When you look at pictures of the **Earth** taken from **space**, you never see the **stars** in the **background**. Why not?

Because cameras don't work in the same way as our eyes. If you were to take a photo of a candle next to a floodlight, you wouldn't see much of the candle because cameras produce pictures based on light gathered in a split second. Our eyes, or more correctly our brains, compile pictures based on signals over a period of time, up to several seconds. So, as we look at the night sky, the longer we look the more light we see, not only because more light reaches our eyes but because our eyes and brains are working together to become better at forming an image. Even so, we still wouldn't be able to see a candle next to a floodlight. And for the same reason, the bright image of the Earth makes it very difficult to see the dim stars behind. Even the surface of the Moon was too bright for stars to appear in the pictures taken by the Apollo astronauts.

Is the **Earth** getting **closer to**, or further from, the **sun?**

The Earth is moving very, very slowly away from the sun, for two reasons. The sun is constantly losing mass because of the solar wind, which is the million kph river of hydrogen and helium nuclei which stream from it. As the mass of the sun decreases, its gravitational pull on the Earth decreases, so the Earth moves slightly further away.

The second reason is to do with tidal forces. In exactly the same way that the Moon is slowly moving away from the Earth, the Earth is very slowly moving away from the sun. In the case of the Earth and Moon, the Moon pulls on the Earth creating tides and slows the Earth's rotation very slightly, making the day longer. This action has a reaction and the Moon's orbit is speeded up. The result of the Moon's increased speed is for its orbit to move outwards, and so the Moon is slowly drifting away from us. But not by much – only 3.8cm per year.

173

The same happens with the sun, but the Earth's influence on the sun is much smaller than the Moon's influence on the Earth. The result is that the Earth's drift away from the sun is very, very small indeed.

How does the **sun burn** if there's no **oxygen** in **space?**

The sort of burning that is happening on the surface of the sun is not the same sort of burning we see in fires on Earth. In a simple fire, oxygen is mixing with a fuel, and heat and light are produced. If there's no oxygen you can't produce a flame.

In the sun, the heat and light comes from a different sort of reaction called nuclear fusion. Here the inner bits of atoms, the nuclei, collide and fuse together. Each fusion reaction releases a million times as much energy as a single chemical reaction, which is how the sun can burn so brightly and for so long. This reaction does not need oxygen so it doesn't matter that there is no oxygen in space. The sun can burn quite happily without it.

How **long** would it take to **drive** to the **sun** in a **car?**

If you can keep going at 70mph, it would take 152 years. Be warned, there are no service areas on the way and

breakdowns are likely. It would take 5 months to reach the Moon at this speed, and 15 days to travel round the world. But it would take 40 million years to reach the nearest star, and even if you swapped your car for a speedier aeroplane it would still take 5 million years to get there.

So how **far** to drive to **space?**

Anyone who has flown higher than 80km can be officially described as an astronaut, so you could argue space is 80km away. The Earth's atmosphere gradually thins the higher you go – it doesn't come to an abrupt end – and up to about 80km high the composition of the atmosphere is about the same as it is on the Earth's surface (i.e. 78 per cent nitrogen and 21 per cent oxygen) although the density of the air gets very much thinner with increasing altitude.

Above this height, the chemical composition changes, and once you reach 2,000km the atmosphere is mainly hydrogen. The outermost limit of the Earth's environment is known as the magnetopause, which lies about 70,000km above the Earth in the direction of the sun, and at least 1,000,000km in the direction away from the sun.

But as far as 'space' is concerned, you could say it starts at 80km up. If you were able to drive a car vertically into the air, it would only take you an hour to get there. And there would be plenty of room to park.

What is the **Moon made** of?

The Moon formed at the same time as the rest of the solar system when it condensed out of a swirling cloud of rocks and gas about 4.5 billion years ago. Many of the planets have their own moons spinning round them, just like Earth, but the Earth's Moon is the biggest in the solar system.

There are a number of theories about how the Moon formed. Scientists used to think that it was a huge piece of rock that had been torn out of the Earth leaving behind a large hole which is now the Pacific Ocean. We now think that is unlikely.

The Moon probably condensed out of those same swirling gases, first as a separate mini-planet, which was then captured by the gravitational pull of the Earth and became our Moon.

Lunar samples brought back by astronauts show that it is made from a volcanic rock called basalt, similar to many of the volcanic rocks on Earth. These form when volcanoes erupt, throwing out molten rocks into the air or the sea which cool very quickly and form rocks.

Inside the Moon there is a crust, a mantle and a core, like the Earth. But the Moon has cooled down a lot more than the Earth and its mantle is no longer molten, which is why there are no active volcanoes on the Moon any more. There are, however, occasional earthquakes, or moonquakes as they should be called.

Why do we always see the
same side of the **Moon?**

The reason our Moon always keeps the same face turned towards us is because the time it takes to orbit the Earth is the same time the Moon takes to rotate about its own axis.

The Moon takes about 27 days to revolve once around the Earth, and takes 27 days to turn once on its own axis. So, in the time the Moon revolves around the Earth it has turned exactly once on its axis – this is called synchronous rotation.

Why does the **Moon** appear **bigger** near the **horizon?**

This is a famous optical illusion, and there's no good explanation for it yet. If you take a ruler and measure the size of the Moon when it is close to the horizon, and then again when it's higher up, you will find that there is no difference. However, your eyes tell a different story and the moon nearer the horizon appears much bigger. There are several theories.

Firstly, when we see things which are on our horizon, such as buildings and trees, we know they are not far away – just a few miles. When we see the Moon on the same horizon our brain thinks the Moon is at that distance too, and so makes it appear bigger. Also, experience tells us that as things pass overhead they are getting closer to us and appear larger. The Moon does not do this because as it passes overhead, it remains at the same distance. The brain compensates for this by making the Moon look smaller as it passes overhead. Interestingly, if you try to capture this enlarged moon by taking a photograph, you will be disappointed because the moon will always appear smaller than the image you remember.

Remember that the moon is not in a perfect orbit around the Earth. At perigee it is nearest to us, and at apogee it is furthest away. When it is closest, it appears to be 1.1 times bigger. But the horizon effect works at both apogee and perigee and so the optical illusion is nothing to do with the moon's non-circular orbit, as is sometimes thought.

Who is the '**Man** on the **Moon**'?

Early in the history of the solar system our part of space was filled with meteorites, asteroids and comets which were frequently in collision with the newly formed planets. Both the Earth and the Moon took a huge number of hits and this covered their surfaces with craters. However, the Earth has volcanoes and weather so, over time, the craters were either filled in or eroded. The Moon, on the other hand, has no climate and no volcanoes, so the craters piled up on top of one another forming a heavily marked surface with basins and mountains.

When the sun shines on the Moon it casts shadows, and since we always see the same side of the Moon we always see the same shadows in the same patterns.

The human brain is programmed to see faces, and it is common to see faces in many different abstract objects. Faces have been seen in pieces of toast, the froth on top of coffee, and in the shadows of clouds on hillsides. The man on the Moon is just another of these. In Chinese tradition, it is not a man on the Moon, but a rabbit and a three-legged toad.

CRASHING COMETS TO GROWTH IN SPACE

In **space**, where does the **heat go?**

Space is a vacuum so no heat can be lost by conduction. Instead, it is lost entirely by radiation. During heat radiation an object gives out infrared radiation which has a longer

wavelength than light. This radiation consists of photons which have energy, so heat energy is taken out of the object reducing its temperature.

From our experience on Earth, losing heat by radiation may seem a very slow way to lose heat, but this is not the case. On Earth everything around us, the walls of your room, the ground, and the air in the atmosphere, is at a similar temperature. This means that you gain nearly as much heat from radiation as you give out so you don't notice the effect. But in space you are surrounded by nothing and the effective temperature of empty space is close to absolute zero (−279°C). So in space you would lose a lot of heat by radiation. But equally the thermal radiation from the sun is strong, so the side of you facing the sun would get very hot − hundreds of degrees − and the side facing away would get very cold. This causes lots of problems for engineers who have to design satellites which can withstand both extremes of temperature.

Why do some **comets** have more than **one tail?**

All comets have more than one tail − it's just that there's only one type of comet tail usually seen in normal light. The bright one you see is formed from ice and dust burnt off the comet's nucleus as it heats up on approaching the sun. As it heats, stresses on its crust create cracks in the surface and jets of fresh ice and dust leak from inside the comet and 'boil off' into space. It is the material from these jets that forms the dust tails we see. The other tail consists of electrically charged particles, called ions, which are often blue and can only be seen under suitable conditions.

But a comet's tail isn't always following straight behind the comet, as you might expect. The dust tail bends in a curve as the comet swings around the sun which makes it appear not to be following the comet in a straight line. The ion tail's shape is controlled by the solar wind – a stream of particles blown out at high speed in every direction from the sun – so the ion tail always runs directly away from the sun.

Did an **asteroid** really kill off the **dinosaurs?**

The dinosaurs were wiped off the face of the Earth 65 million years ago, somewhere between the Cretaceous era and the first age of mammals called the Tertiary era. All we know is that an asteroid *did* collide with Earth at the so-called Cretaceous/Tertiary boundary and geologists have now identified what they think is the crater left by the

asteroid's impact – the Chicxulub crater in the Gulf of Mexico. A large asteroid colliding with the Earth would have sent shockwaves around the planet and created tidal waves, throwing vast amounts of fine dust into the upper atmosphere where it could have stayed for many years. This would have blocked out the sun, cooling the Earth to a temperature at which the dinosaurs could not survive. However, while some scientists support this theory, others believe that the dinosaurs died out over a much longer period than just a few years, and say the asteroid collision was just a coincidence, or just one of a number of contributing factors.

There is another theory. Some scientists suggest that a dark, mystery object, nicknamed 'Nemesis', might orbit the sun at a huge distance and regularly send comets plunging into the inner solar system where they might collide with planets. Another suggestion was a comet bombardment related to the sun's movement in the galaxy. All of these theories have yet to be proved.

What happened on **Jupiter** after the **comet crash?**

One of the biggest events in recent astronomy was the collision between Comet Shoemaker-Levy and the planet Jupiter in July 1994. The explosions, as fragments of the comet blew up in Jupiter's atmosphere, were the most powerful ever seen in the solar system, and it served as a reminder that the threat of a similar impact on Earth should be studied and prepared for.

Astronomers watched with interest as the newly discovered comet first passed close to Jupiter in 1992. To their amazement, Jupiter's gravity changed the comet's orbit, breaking it up into a string of smaller fragments and putting them into a collision path with Jupiter. Unfortunately, the region where the impact happened was on the far side of Jupiter as seen from Earth, but astronomers were able to watch as the comet fragments headed towards Jupiter at over 210,000kph. Many of the impacts happened just behind the limb (the edge of the planet's disc), so although they could not see the impact itself, astronomers could see huge glowing gas clouds shooting into space as the fragments exploded. Unlike Earth though, Jupiter is so huge (136,000km across) that it was able to soak up such a massive impact.

Although Jupiter is nearly all gas with only a tiny solid core, the speed of the fragments' passage turned them into giant shooting stars which heated as they collided with the thickening atmosphere, finally exploding. By the time Jupiter's spin had brought the crash sites round so that we could see them from Earth, they showed up as a series of dark bruises in the clouds caused by darker-coloured gases being stirred up inside Jupiter.

After seeing the power of the planet's gravity in operation for the first time, some astronomers have suggested that Jupiter acts as a 'guardian angel' for life on Earth, soaking up or disrupting the orbits of many large comets that enter the inner solar system, making it less likely that the Earth will suffer a catastrophic impact. Jupiter could be Earth's best friend.

What is the **danger** of a comet or asteroid **hitting** the **Earth?**

In the long term, it's an absolute certainty. The Earth is already covered with the signs of past impacts. We now know for certain that a large object hit Earth at the same time as the dinosaurs were wiped out (see above) and this was only one impact of many. As recently as 1908, something (probably a comet fragment) crashed into Siberia at Tunguska, wiping out 2,000 square kilometres of forest. Had it struck the Earth only two hours later, it would have wiped out Moscow.

There's always a chance of a comet appearing from the outer solar system on a collision course, but the real problem is that the inner solar system is much more crowded than we first thought. Many scientists are now turning their attention to this problem and have developed 'Project Spaceguard' – a plan to map the orbits of every substantial object with an earth-crossing orbit. With a network of powerful telescopes dedicated to the search, they predict Spaceguard could increase the discovery rate of these objects to several thousand per year.

If an approaching object was discovered in time, then it might be possible to prevent it. An obvious way would be to destroy the object with a rocket-delivered nuclear bomb. However, such an explosion could multiply the problem, creating a swarm of smaller planetoids on course for Earth. A far better solution is to divert the asteroid. Even a small nudge in the right direction could turn an impact into a near miss, and diverting the object does not require nearly as much energy as destroying it. Diverting the asteroid could be done in one of two ways. A small nuclear bomb landed on one side of it and then detonated would do the trick, or a powerful rocket engine anchored to the surface of the asteroid, pushing it slowly into a different course over a long period, could also alter its course.

Are we in any **danger** from the **asteroids?**

The first thing to say is don't panic! It's very unlikely that you will be struck by an asteroid. About 80 asteroids are known to cross the Earth's orbit, 7 of which are called Aten asteroids: these regularly transit our orbit. The Aten asteroids are typically less than 2km across, and do come relatively close to us. The closest passed about 690,000km from the Earth in 1989 – less than twice the Moon's distance. This was thought to be a bit too close for comfort.

It has been calculated that 5 strikes may occur in a million years, which is not a very long period of time geologically speaking. Comets could also be a problem. Each year approximately 30 comets pass inside the orbit of the Earth with a collision rate of one every 10 million years. Comets can be up to 2km across and if they struck the Earth would cause a crater 30km in diameter, which is roughly the size of London.

A major problem would be the amount of dust thrown up from such an impact. This would darken the sky and reduce the amount of energy and light from the sun causing the planet to cool. The chances are still slim, and if we were threatened it's most likely that we would be able to take action to deflect the incoming intruder (see above).

Is it true that I'm more likely to get **killed** by a **meteor** than by a **flood** or an **earthquake?**

No one has ever reported being hit by something from space, but I suppose they probably wouldn't be around to tell the tale. There have been reports of a car being hit and a cow being struck down by something extra-terrestrial. But the chances are very small indeed.

Four-and-a-half billion years after the formation of the solar system, a great deal of the rubbish that was left over has already collided with the nine planets, which means there's not much left to hit the Earth. At least, not much that's big enough to make it through the atmosphere and actually hit the Earth's surface. If you take the number of meteorite impacts every year, then compare the size of the Earth, and then *your* size in relation to the Earth, the chances of one of those meteorites hitting you are perhaps one in a trillion.

The risk of being killed by an earthquake or a flood depends largely on where you live. For example, living in London you are unlikely to die in an earthquake – the risk is almost zero. However, in an earthquake zone like Colombia, 2,000 people died out of a population of about 35 million in a recent quake, so the risk of death was around 1 in 17,500.

For floods, again the risk of death depends on where you live. Living somewhere like Bangladesh, which is prone to devastating floods, the risk of death is obviously high. The floods of 1998 resulted in about 2,000 deaths from a population of 120 million. So the risk of death was around 1 in 60,000.

You can see that if you live in an earthquake or flood-prone area, then your risk of death from earthquakes or floods is much higher than being hit by a meteorite. However, if you live somewhere like London where the risk of death from earthquake or flood is so small, it may well be true that you are more likely to die from being hit by a meteorite.

How **far** can a **telescope** see?

The telescope that can see the furthest into space is the Hubble Space Telescope which was launched from the Space Shuttle Discovery in 1990 and orbits the Earth above the atmosphere. It can see objects up to 11 billion light years away. How far is this? Well, one light year is the distance light travels in one year which is about 6 million, million miles. But Hubble is not the largest telescope built; that is to be found on the island of Hawaii and is called the Keck Telescope and its mirror is 10m in diameter. The Hubble is also a reflecting telescope but, although its mirror is only 2.4m in diameter, because the Earth's atmosphere is between the Keck Telescope and the stars it is looking at, Hubble can see further.

Do all the planets **spin** in the same **direction?** How **fast** do they all spin?

The Earth spins in an anti-clockwise direction seen from the North Pole. Most of the other planets spin this way too. Venus, however, spins in a clockwise direction, or in a 'retrograde' way, which means that on Venus the sun rises in the west and sets in the east. Why it should do this is not really known. It's possible that Venus was hit by a large object early on in its formation that sent it spinning in the opposite direction.

The planet which spins the fastest is Jupiter – it takes just under 10 hours to spin completely. The Earth, as we know, takes about 24 hours to rotate fully. Venus takes the longest time to spin at a massive 243 days. Mercury, which is the planet nearest to the sun, takes just 59 days. Here are some others: Mars – 24 hours 27 minutes, Saturn – 10 hours 39 minutes, Uranus – 17 hours 54 minutes, Neptune – 19 hours 12 minutes.

What are the most **powerful** rockets?

The most powerful rockets ever built were the American Saturn Vs – the huge rockets which took the Apollo missions to the Moon. A Saturn V rocket stood 111m tall on the launch pad and weighed 3,000 tons with a full load of fuel. The payload – which is the manned capsule – was contained in a tiny section at the top of the rocket.

Saturn Vs took all their power from liquid fuel – hydrogen and oxygen burning together. At the time, solid-fuelled rockets were only just being developed, and so there wasn't any other option. The Saturn Vs were built as a series of three ascending stages, each with its own engines and fuel supply.

The largest stage was at the bottom, the smallest at the top directly underneath the manned capsule.

Launching a space probe out of the Earth's gravity requires much more energy than putting a satellite into orbit. All planets are surrounded by a large gravitational field which gradually becomes weaker at greater distances. Escaping from this is a little like climbing up a hill that's steep at the bottom, but gets less steep towards the top. At any time, gravity can pull the spacecraft back to Earth but things get easier the further it goes. If it can reach the Earth's escape velocity – the speed at which an object must be propelled away from a body so that it is no longer under gravitational attraction – then the spacecraft will be travelling so fast that gravity will never be able to slow it down and recapture it. Escape velocity is a very high speed – 11.6km per second.

This speed was achieved by the largest stage of the Saturn V and once it exhausted its fuel it was jettisoned and fell back to Earth. Then the next stage started giving the rocket more and more speed. Once all the stages were used up, the case around the payload fell away and it effectively drifted (only at very high speed) towards its target.

It's a mistake to think that a spacecraft goes all the way to its destination with a rocket burning behind it. Once the spacecraft is on its way, there is no friction in space to slow it down, so once the gravity of its home planet has been overcome it will keep moving until it hits something, or is deliberately slowed down, usually with small 'retrorockets'.

A Saturn V could have put objects weighing up to 140 tons into a low Earth orbit (400km high), but when aiming for the moon in the Apollo missions the manned capsules were limited to just a few tons. Saturn V was a very expensive rocket, and wasteful. As each rocket stage fell back to Earth, it was destroyed or lost at sea, and couldn't be retrieved and reused like a lot of the parts on today's rockets.

Why doesn't a **spacecraft** burn up when **taking off**, just as it does on **re-entry?**

A spacecraft burns up on re-entry to the atmosphere as a result of its high speed, which causes friction between the craft and the atmosphere. But in order for sufficient heat to be produced, the spacecraft must be travelling at a very high speed. When it takes off, it simply isn't going fast enough for this to occur. As the spacecraft climbs it starts to travel at a higher speed, but the atmosphere is becoming less dense the higher the rocket goes, so heating doesn't happen to a significant degree.

On re-entry, a spacecraft is moving much faster as it rubs against the atmosphere, creating more heat. Re-entry speeds can be as high as 10km per second.

Do people **grow** in **space?**

In space, the body does not experience the same downward pull of gravity that it does on Earth. The result is that the gaps between the vertebra in the spine expand and astronauts do get slightly taller. When they return to Earth, the gravity quickly squashes them back to their normal height.

6

CAN
You Just
Explain . . .

Yo-yos to Frisbees

Clouds to Vapour Trails

And All the Other Things I Don't Understand ...

YO-YOS TO FRISBEES

Why does a **yo-yo** come back up the **string?**

A proper yo-yo isn't fixed to the end of the string – it sits in a loop. As you will find if you drop it, and don't jerk your hand upwards at the right time, it will spin at the bottom until it runs out of energy.

So why does a yo-yo come back up a string when you jerk your hand? Imagine you're looking at the yo-yo side on, so that the yo-yo is spinning

anti-clockwise. It spins in the loop of string because of the angular momentum you have given it by releasing it and letting it drop. If you stayed perfectly still the yo-yo would sit there, spinning in the loop till the small amount of friction eventually overcame the angular momentum and the yo-yo came to a halt.

But when you give it a jerk, you momentarily increase the friction between the string and the yo-yo, it 'catches' and the remaining angular momentum enables the yo-yo to wind itself back up the string. As the yo-yo rises, its angular momentum decreases until it comes to a stop, when it will fall again. It may then rise a short way up the string if you give it another jerk, but by now the angular momentum will be rapidly decreasing (unless you've given it another flick) and it will soon come to a halt. However, if you keep giving the yo-yo energy, it will keep going up and down that string for as long as you go on flicking it.

Why does a **football swerve?**

A football will swerve only if it's spinning. Newton's first law of motion says that any body travelling through space will keep going at a constant speed and direction unless a force acts upon it. In the case of the football, it swerves because a force *is* pushing it sideways.

Imagine the spinning ball flying through the air, and let us assume that we are looking at the ball from above. The ball spins clockwise, so relative to the ball its left-hand side is moving towards the front of the ball while its right-hand side is moving towards the back of the ball. This means that the air flow on the left is slowed down on the surface of the ball, and the air flow on the right is speeded up. This is the source of the force which causes a spinning ball to swerve.

When air moves faster the space between molecules is very slightly increased, which means a decrease in the pressure at that point. A decrease in pressure is equivalent to a 'sucking' force, and this is what is experienced by the right-hand side of the ball. Similarly, when the air is slowed down on the left-hand side of the ball, an increase in pressure results in a 'pushing' force. Overall, the ball will start to swerve towards the right. If the ball is spinning the other way, it will swerve to the left.

How can you **skim** a stone across a **pond?**

In order to skim a stone it must be spinning and tilted slightly so that its leading edge is a bit higher than its trailing edge, in order that the trailing edge will hit the water first. But even though it has hit the water, the stone wants to keep spinning along the same axis, like a gyroscope. So the whole stone skips up from the surface of the water, flies a little distance, and then repeats the process. It will keep jumping as long as the stone is spinning. Gyroscopes always try to keep spinning in the same direction, so rather than flipping over when the stone hits the water, it jumps in order to keep its axis pointing in the same direction.

You can't skim a spherical stone because there is no trailing edge to hit the water; and you can't skim a stone that isn't spinning because it would simply sink.

When a **spinning top** is spinning will it **weigh** the same?

A few years ago there was a report of a spinning gyroscope that appeared to weigh less when it was spinning. However, the laws of physics say that this shouldn't be the case and no one has since been able to repeat the experiment or explain how it might have happened. So, we can probably safely say that a spinning top will weigh exactly the same whether it's spinning or not.

When something spins, it doesn't generate any force in the upwards or downwards direction. For something to move upwards, and therefore weigh less, there has to be an upwards force and such a force isn't generated by a spinning top.

200

Why does **spinning** make a **frisbee** stable in **flight?**

It's the same thing that makes a moving bicycle more stable than a stationary one, and stops a spinning top from falling over – angular momentum, which all spinning things have.

But frisbees are carefully shaped, which is also important in flight. If you look at the rounded edges of a frisbee you'll see that they look very similar to the leading edge of a wing. When the frisbee passes through the air, lift is created in the same way as in an aeroplane, and that helps to keep it elevated. But if the frisbee has no spin it has no stability, and will soon wobble and crash to the ground. By spinning, the frisbee has angular momentum and gains stability in the same way as a spinning bicycle wheel, until the drag effects of the air overcome it. And that is when the frisbee falls to Earth.

CLOUDS TO VAPOUR TRAILS

If you were in a **cloud**, would you **drown?**

You've probably been in a cloud already so you can guess the answer. Fog and mist are nothing more than low-lying clouds, and it's perfectly easy to breathe in them.

A cloud is a vast collection of extremely small drops of water floating in the sky. But that's not the same as being made entirely of water. There is plenty of air between the drops, so you would still be able to breathe. In very thick clouds you would get quite wet, but you wouldn't drown.

How does the **air** get under **frozen puddles?**

Any air that is under the ice will find the highest point and collect there. The question is, where does the air come from?

There are two sources. Firstly, as the underside of the ice melts any air in the ice is released. Secondly, because the water takes up less room than the ice, an area of low pressure is formed where the ice is melting. This pulls the ice in and can crack it, even slightly, to allow air in. So some of the air has emerged from the ice, and some may have been sucked in as the ice melts.

Why does **freezing** water **expand?**

Water is certainly strange stuff. Most liquids which turn into solids as they cool tend to contract as part of the cooling process. But as water reaches the point where it becomes like a slushy drink (just before it turns into solid ice) it becomes less dense. The magic figure is 4°C, and up to that point the water has behaved normally and steadily contracted as it cooled. But when it hits 4°C it starts to expand until the most dramatic expansion (about 9 per cent) which occurs at 0°C when it turns to ice. Water is the only substance where the solid is less dense than the liquid. If it weren't, then ice cubes would sink to the bottom of your drink.

Each water molecule consists of 2 hydrogen atoms bonded to 1 oxygen atom by hydrogen bonding, which is a form of bonding that occurs between hydrogen in

I molecule and a negative ion in another molecule. In liquid water, as the molecules move freely past each other, bonds are formed and fractured quite easily. But by the time water has cooled to 4°C, the molecules' energy has dropped sufficiently for movement to slow, so that each H_2O molecule forms more stable hydrogen bonds with up to 4 fellow molecules.

At freezing point, the H_2O molecules are lined up in a frozen crystal lattice and the molecules held rigidly apart. That means more empty space between molecules, so the frozen water occupies more room than the water from which it came.

When you stir a glass of **water**, the sound made by **banging** the top of the **glass** goes up. Why?

When you bang the side of a glass, the noise it makes depends on the length of glass that is free to vibrate. The higher up the sides of the glass the water comes, the shorter the vibrations, and the shorter the vibrations the higher the note – just like plucking a string on a guitar. As you stir a glass of water, the water is forced up the sides of the glass making the sides shorter and the note higher.

What happens to a **ping-pong** ball on the surface of a **bucket** of water in an **elevator?**

Just as bits of your body wobble about in an elevator as it speeds up and down, so too would a ping-pong ball floating on water. As the elevator goes up, the ping-pong ball will bob down in the water fractionally, and then return to its normal position, and as the elevator comes to a halt the ball will continue upwards slightly and then return to its starting position. The opposite would happen as the elevator went down. The question is, why?

It's all to do with inertia. The inertia of the ball is its resistance to a change of velocity. So, if something is stationary it has a certain resistance to any force that tries to make it move, while anything that's moving has a certain resistance to any force that tries to stop it. Because the ping-pong ball isn't physically attached to the elevator, when the elevator starts to move the ping-pong ball gets left behind. But because it's floating on water, which naturally tries to keep its shape, the ping-pong ball gets pushed or pulled back by the water to where it was before.

205

Why do the **particles** of solids, liquids and gases **behave differently?**

The particles which make up a solid cannot move freely since they are packed very closely together and can only vibrate about fixed points. The more a solid is heated, the faster the particles inside it vibrate until eventually they break away from their fixed points and the solid melts.

The particles in a liquid are not quite as close together, and they can move freely anywhere within their container. If the liquid is heated, the particles inside it move faster and faster until they start to break free from the surface and the liquid evaporates.

The particles in a gas are usually spaced quite far apart compared to liquids and solids. Gas particles can move freely anywhere within their container and their average speed is much higher than the particles in a liquid. There are forces of attraction between all particles. These are strongest in solids and weakest in gases.

Why do you see **trails** of vapour on the **edges** of **aeroplane wings?**

The trails often seen during landing and take-off over the upper surface and trailing tip of an aircraft's wing are caused by the low pressure created by the shape of the wing. Air at low pressure cannot hold water moisture as effectively as air under higher pressure, so any water vapour tends to condense out and appears as visible trails of vapour. You can see them best when a plane lands in air which is very damp.

The broad vapour trails which criss-cross the sky are different. These are caused partly by fine water vapour from the exhaust which crystallises in the low temperature conditions. Aircraft put out quite a lot of water in their exhaust – every gallon of fuel burned can produce a gallon of water. The drier and colder the air, the more prominent the trail. As the hot exhaust gases hit the cold air they rise rapidly and cool so quickly that the water vapour soon condenses and freezes. These ice crystals act as a nucleus on which more moisture is deposited. The trails are often called contrails – condensation trails.

AND ALL THE OTHER THINGS I DON'T UNDERSTAND ...

If you eat a bit of **chocolate** and there's some **foil** left on, why do your **teeth hurt?**

It will only hurt if you have fillings made of one kind of metal, and if the foil is made of a different kind – a highly likely combination. What you are sensing is galvanic action between the two. This is the creation of an electrical current between two dissimilar metals when they come into contact.

Galvanic action occurs because different types of metal have a different likelihood of becoming ions, which are electrically charged atoms. In order for atoms in the metal to become ions there must be electrons available, usually from a solution such as water or, in the case of your mouth, saliva. So, if you have two metals – each with a different likelihood of becoming ions – and saliva in between, electrons will move from one to the other.

The movement of electrons is the same thing as saying that an electric current is flowing, so you get a very small electric shock which sets off the nerve endings in your teeth. This is what hurts.

Why are bananas **curved?**

Bananas grow on tall trees, in big bunches around the top of the trunk. There could be more than fifty bananas in a bunch all packed tightly together. The bananas are curved so that they can fit closely together allowing more bananas to grow around the tree. If you look at a bunch of bananas, you will see that they fit together almost like the fingers of a hand.

Do **computers** have **feelings?**

No. To experience pleasure or pain you require a consciousness, which computers don't have. No computer has ever achieved consciousness. However, in a lab in San Francisco scientists have embarked on a project to create a computer/robot baby which has been programmed to want stimulation and attention, like a human child. It has been equipped with eyes (a camera) and touch (pressure pads) and is hooked up to a very complex computer. Is there enough feedback from these sensory organs for it to begin to show emotions? So far all it does is follow people around the lab. Still early days yet.

How high could a **skyscraper** be built, using present **technology**, before it **collapses** under its own **weight?**

The weight of a building really has nothing to do with its ability to stay upright. Even the lightest structure, like a house of cards, will collapse if it is not properly reinforced. So there is no engineering reason why skyscrapers shouldn't get taller and taller. The only thing that limits building height is money and practicality.

Imagine some of the problems with a hugely tall structure: how long would it take to leave the building? The air pressure at the bottom of the building would be greater than at the top – a bit like a mountain – so what effect would that have on people going to work there? Would the upper floors have to be pressurised like a high-flying aeroplane? These are just a few of the reasons why skyscrapers will only get so tall.

Why do razor **blades** become **blunt?** After all, **hair** is pretty **soft** stuff.

The fact that steel is much stronger than hair doesn't make much difference. Water is much 'softer' than rock, and yet the majority of the Earth's surface has been moulded by the action of water.

A blade is very thin – the thinner the blade, the sharper the cut. And the thinner the blade, the fewer atoms there are at the cutting edge. As the blade passes over the stubble on a chin, the hair knocks individual atoms off the steel blade, and since there are many thousands of hairs it doesn't take long to slightly dull the blade. Even if each hair only removes one atom, the blade can be dulled quickly.

All blades will go dull eventually simply by the repeated action of the things they come into contact with, no matter how hard or soft they are. Cut enough ripe bananas and you'll blunt a kitchen knife.

What is it that comes out of **radioactive** materials?

The answer is energy, and it emerges in three different forms – alpha, beta and gamma rays. When radiation was first discovered nobody knew what it was, but they were sure there were three different types so they called them alpha, beta, gamma – the first three letters of the Greek alphabet.

Alpha rays are heavy, fast-moving particles with a positive charge which only travel a few centimetres in air and are stopped by a sheet of paper. They are the nuclei of helium atoms.

Beta rays are also particles, but very much lighter and faster-moving than alpha particles. They can travel through a metre or so of air, but are stopped by a few millimetres of aluminium. They have a negative charge and are electrons.

Gamma rays are electromagnetic waves, part of the electromagnetic spectrum like light and radio waves. They have a very short wavelength and are similar to X-rays, but with shorter wavelengths and more energy. They can pass through thick sheets of lead. All three come from the nucleus of the atom. Some radioactive atoms give out alphas and some betas. In some cases an atom will emit a gamma ray when it is settling down after emitting an alpha or a beta – a sort of nuclear burp.

Is *everything*
radioactive?

It is true that there is always very weak radiation around us and it is called background radiation. It even makes a very small number of people ill. The main radioactive elements are uranium and thorium, but there are others. For example, potassium contains a trace of radioactive potassium-40, and all living things contain a small proportion of radioactive carbon-14.

In the nineteenth century, craftsmen used uranium to give glass a nice yellow colour, but had no idea about radioactivity or its dangers. Nor did they realise that luminous paint was radioactive. As a result, workers at an American plant producing luminous dials innocently put their brushes in their mouths and some developed jaw cancer. When above-ground nuclear bomb tests were common, traces of radioactive fallout were everywhere. Strontium-90 was found in children's bones. The accident at Chernobyl has made parts of the surrounding area uninhabitable, but it also had an effect as far away as Wales and the mutton from sheep eating radioactive grass could not be sold. So there's radioactivity almost everywhere, but that doesn't mean *everything* is radioactive.

The French physicist Henri Becquerel discovered radioactivity in 1896 and gave his name to a unit of measurement, called the Becquerel, which is defined as one radioactive disintegration per second. If you were to measure the radioactivity in a loaf of bread it would be 70Bq, an adult person 3,000Bq, a kilo of tea 430Bq, and a kilo of coffee 1,640Bq.

Could a **coin** dropped from the **Eiffel tower** hurt someone, and how **fast** would it be **travelling** when it **hit** them?

A coin dropped from the Eiffel tower would reach terminal velocity – the speed at which the drag of the atmosphere equals the pull of gravity – after about 350m. But as the Eiffel tower is only 320m high, it might be moving a little below that terminal velocity. So, we can assume it would be travelling at something just under 80m/second, which is 179mph – certainly enough to do serious damage.

If you fire a **bullet** straight up, where will it **land?**

On your head. Even though the Earth is rotating, so is the gun, the bullet, the atmosphere, and the person who fired the shot. So, assuming there is no breeze, and that the shot is fired precisely in a vertical direction, and you don't move a fraction, the bullet will theoretically end up back where it started. In practice, it is bound to be deflected one way or another by air movements.

215

If lots of **people** join their **hands** together and **jump** from an **aeroplane**, can you land without **parachutes?**

There are two factors which decide how fast objects fall through air – the mass of the object and its cross-sectional area. If you double the mass and double the cross-sectional area, then the speed the object falls at remains unchanged. But if you double the mass and halve the area, then the speed goes up (and vice versa).

If a large number of people jump out of an aeroplane, then their combined mass goes up. But if they're joined together in the typical formation (each person holding on to the wrist of the person next to them) then the cross-sectional area also increases. What is strange, though, is that because of turbulence, the effective cross-sectional area of all of them is greater than the cross-sectional area of each individual added together. So if you had 100 people with an area of say half a square metre, you'd expect their total cross-sectional area to be 50 square metres. But when you're in formation the cross-sectional area is more like 75 square metres. So the mass goes up, but the cross-sectional area goes up even more. This means the terminal velocity of the group goes down and a large group of people would fall more slowly than a small group!

In a sense, the people are making their own parachute because they are spread out so much. But to answer your question, you could do a calculation which would tell you how many people needed to be joined together to create the same drag as a parachute. No one has yet tried this for real.

Who decides when to **add leap** seconds to the **years?**

How long a second is, and how many seconds make an hour has always been based on the rotation of the Earth on its axis and around the sun. Over the centuries our measurements of time and how the Earth rotates have steadily improved and we came to realise that there are irregularities in the Earth's rotation. Our definition of a second, based on the rotation of the Earth, meant that each second had to vary slightly in length to average out at a standard length for the year.

But in the 1950s an atomic clock was devised that measured seconds not in terms of the rotation of the Earth, but by the number of vibrations of a particular caesium atom. Because this was independent of the rotation of the Earth, each second was exactly as long as the next.

Unfortunately, the most useful type of time is the heavenly one as all navigation is based on time and its links with the stars and planets. So atomic time, although exceedingly accurate, was not very useful in daily life.

In 1972 a new Coordinated Universal Time scale was devised that combined the accuracy of the atomic clock and the usefulness of the heavenly clock. Each second is the same length as a second of atomic time, and to make atomic time and astronomical time match up, leap seconds are added either at the end of the year, or in the last minute of June.

Each year at these times, the International Earth Rotation Service in Paris decides whether a leap second is needed or not, and then notifies the world. The extra second is added at the same time, the whole world over.

Do **alloy** wheels make **cars** go **faster?**

Bad news – alloy wheels have no effect on the performance of a car at all. They are used simply because they are light and strong. The lightness of the alloy means the wheels weigh far less than a same-size steel wheel. By reducing the weight of the wheels, other parts of the car can be remodelled leaving the overall weight the same. For example, the body of the car can be made bigger so that the car has more room inside, but it wouldn't be any heavier.

This is important in terms of cornering. If the car is too light the car will lift off the road. If the car is too heavy then the car will be sluggish. However, since the weight saved by using alloy wheels is usually used up somewhere else on the car, having alloy wheels doesn't make any difference. Alloy wheels are mostly put on to cars to make them look good. They have no effect on the performance of the car at all.

Is it more **efficient** to climb **stairs** two at a **time?**

From a purely scientific point of view, a constant amount of work is required to raise a body by a given height. But what if the body is human and the route is up stairs? Given that muscles are not 100 per cent efficient, can significant savings in energy be made by halving the number of muscle motions and climbing the stairs two at a time? Or is the saving so small that it is swamped by the extra energy needed to make the longer stride?

There seems to be no precise answer because there are too many variables. The economy of stair climbing would depend upon stair height, body mass, leg strength and leg length. However, the amount of work that you would perform would remain pretty much the same because the start point and the end point are the same whichever way you get there. If you jump off the roof of a high building, or go down via the stairs, the work load is the same because the work formula includes overall distance moved, not how you get there.

It's one of those cases where science says one thing, and experience says another.

7

AND

Now the BIG Ones

Lumps of Light to Absolute Zero

LUMPS OF LIGHT TO ABSOLUTE ZERO

What are **quanta?**

Quanta are 'lumps' of light, and they were first described in 1901 by the German physicist Max Planck. He was studying the radiation from black bodies and came to the conclusion that electromagnetic energy is emitted in packets. This idea was the basis of one of the most fundamental theories in physics – quantum theory. When a light source gives out energy, it loses this energy in indivisible packets called quanta. Einstein went further. He said that this energy was converted into separate packages of energy called photons which carry the energy of light and other electromagnetic radiation.

Every form of energy comes in 'bits', or quanta. Imagine you have a lamp at home and connect it to a dimmer, you might think that when you turn the knob the fade is smooth and continuous. Actually, you'd be wrong. Your eyes aren't sensitive enough to detect it but the decrease in brightness comes in steps. There's no such thing as a truly smooth fade because energy comes in chunks. The world is digital not analogue.

In the case of your lamp: say it loses 100 quanta a second, which it does by emitting 100 photons a second (this is just for the sake of argument – 100 photons a second is so dim you probably wouldn't see anything). Now, as you turn the dimmer switch, the light fades because it is losing fewer quanta a second and so giving out fewer photons. As you turn the dimmer, the number of photons is gradually reduced until you come to the very last one, and then the light goes out. The whole point of this exercise is to show that energy is not smooth, it comes in packets, and these are called quanta.

What is **Schrödinger's** cat?

This is the name given to one of science's most famous philosophical problems, and it is connected with a whole branch of physics called quantum mechanics.

Quantum mechanics tries to explain how fundamental particles, such as electrons, interact with each other, and it contains some apparently odd ideas. One of these is that you can't measure the properties of these fundamental particles *exactly*. When you measure the position of a particle, the simple act of measuring it disturbs the particle and alters what you are measuring. Take time to think about that and you'll understand why quantum mechanics sits on the boundaries of science and philosophy.

This draws attention to how absurd quantum mechanics seems when it is applied not to fundamental particles, but to macroscopic (big) things like you or me, or cats. The problem goes like this. Imagine you had a radioactive atom.

Radioactive atoms have surplus energy and are unstable. At any moment they will give out this surplus energy and return to being a normal atom. There is no law in physics which allows us to say EXACTLY when this will happen. All we know is the PROBABILITY of it happening at a given time. We say it is in two 'states' – one excited and one unexcited. According to quantum mechanics as soon as we measure the atom it will definitely be in one state or the other, but until then the atom is in-between, it is in a 'superposition' of each state – that is, it's a bit of both. This is fine when dealing with things like atoms, which are small and so obey the rules of quantum mechanics, but what about big things?

Schrodinger's cat

One day, a guy called Erwin Schrödinger came up with an idea for an experiment. Put a cat into a box (he said) with a fragile bottle of a deadly poison, a hammer and a radioactive atom. If the atom decays then a mechanism detects this and swings the hammer, breaks the bottle of poison and the cat dies. If the atom doesn't decay the mechanism doesn't move the hammer, the poison stays in the bottle and the cat lives. As before, until we open the box to measure the atom we don't know which state the atom is in, so it must be in a mixture of the two. But, and this is the whole point of the problem, what has happened to the cat? Is the cat, like the atom, in a mixture of states – both dead and alive?

To sum up: an observation or measurement itself affects the outcome, so you can never know what the outcome would have been if you'd never looked.

Since Schrödinger posed this question in 1935, nobody has come up with a satisfactory answer. But don't worry, nobody is going to try this experiment with a real cat. The point is not whether the cat is alive or dead when you *open* the box but what is happening to the cat while the box is closed, so doing the experiment for real wouldn't be any help.

I am pleased to announce that no cats have lost their lives in the course of answering this question. Incidentally, Schrödinger himself was so perplexed by the issues raised by this, that he was reported as saying that he wished he'd never thought of it in the first place.

If a car **travels** at 1,000mph with its **headlights** turned on, does the **light** travel **faster?**

If a man in the car fired a bullet forward at 500mph, and the car was doing 1,000mph, then the bullet would travel at 1,500mph. But if he switched on the headlights, the light would travel at its usual speed of 186,000 miles per second, not 186,000 miles per second plus 1,000mph.

Light is peculiar in that it travels at the same speed for everybody in the universe no matter how fast the light source is travelling. The speed of light and the laws governing it were laid down by James Maxwell, a nineteenth-century Scottish physicist. Maxwell's mathematics did not, however, state for whom this speed of light is applicable – the man in the car or the man on the pavement. It was later realised that it is correct for everybody.

If you **threw** something out of the back of a **spacecraft**, would the **object** be attracted by the **gravity** of the ship and follow along **behind it?**

No, because Newton's second law of motion says that anything which starts moving in a straight line will continue to

227

do so if it is not subjected to other forces, such as friction. In space there is no friction to slow objects down, so anything which is launched out of a spacecraft would have some propulsion of its own and would therefore continue to travel in the direction it was sent. Any gravitational force exerted by the craft would be negligible, especially by comparison with the gravitational fields of the suns or stars in the vicinity.

What causes an **atom** to **decay?**

Atoms lose energy in order to become more stable, but it is impossible to predict when it will happen. We can predict, say, that half the atoms in a sample of radium-226 will decay in the next 1,620 years, but we cannot say which half. A particular atom might decay in one second, or it might wait 3,000 years. Even if we knew exactly what was going on inside a nucleus, we still could not predict when it was going to decay. The decay of an atom is completely random – it is an event without a cause.

This is easy to prove if you set up a Geiger counter (which measures radioactive decay) close to a radioactive source. Measure the counts over a period of 1 minute, and do this 100 times. Calculate the average. Repeat the experiment. The two averages will be very close to one another, but the individual counts will vary quite a lot because the individual decays cannot be predicted.

What do **scientists** mean by '**splitting** the **atom**'?

'Splitting the atom' is another name for nuclear fission. The word fission was first used in biology to describe what happens when the nucleus of a cell splits and one cell becomes two. Scientists who first studied atomic nuclei were reminded of what happened in biology, and used the same word.

Fission occurs when the nucleus of an atom splits into several fragments, and two or three neutrons are emitted.

This can lead to a chain reaction as the released neutrons go on to create fission within other nuclei, releasing more neutrons to 'divide' more nuclei. If you control this process, you have what is called a nuclear reactor which you can use to produce energy. If the chain reaction goes unchecked, you have a bomb on your hands.

The atom was first split in the laboratory in 1915 by Ernest Rutherford who, with Niels Bohr, first proved the idea that an atom consisted of a positively charged nucleus circled by electrons. Rutherford fired alpha particles into atoms of nitrogen and found that occasionally the nucleus of the atom would split and a small amount of hydrogen was produced. He thus became the first person to 'change' one element into another artificially.

If **light** doesn't **weigh** anything, how can it be **bent** by a **prism?**

The amount light bends depends on the atoms it collides with as it travels.

Light is a wave, and when that wave meets an atom it causes the atom to wobble in its wake, just as a ripple in a pond can bob a floating stick up and down. But light doesn't move the whole atom – the nucleus stays where it is and the cloud of electrons moves from side to side.

230

The cloud of electrons takes energy from the light and the light is 'held up' on its journey through the glass until the electron cloud has returned to its starting position. If the electron cloud is pulled off centre a long way, then it takes a long time to return to its normal position and the light is slowed down a great deal.

Higher frequency light, towards the bluer end of the spectrum, has more energy, so it can pull the electron cloud around more but pays the price because it is slowed down more on its travels. So the higher frequency light is slowed down more by the glass, the lower frequency light slowed down less. This is what makes it bend and gives you a rainbow of colours.

As everything is made of **atoms**, how can things be **transparent?**

You have to look at what makes objects opaque to answer this. Everything is made of atoms joined to other atoms by bonds, and these groups of atoms vibrate at a certain frequency. When light falls on them, some of the energy of the light is absorbed. This energises the atoms and electrons around the atoms so that they're in a higher state of 'excitement', if you like. But they can't stay this excited for long (who can!) so they quickly drop down to their normal, more relaxed state, getting rid of that excess energy as they do so. This energy is given out as light, and the colour of the light depends on the amount of energy. So, the colour of an object depends on how the light falling on it interacts with the atoms and electrons within it. Blue objects have atoms that are great at absorbing, and then re-emitting, blue light, and white objects are great at absorbing and reflecting all the light that falls on them.

Transparent objects are simply those that don't affect the light that falls on them at all. It passes through the material without being reflected.

Is there a maximum **temperature** in the **universe?**

The temperature of an object is a measure of the energy of the atoms or molecules in that object. You could argue that the maximum temperature is infinite because we can continually increase the amount of energy of any individual molecule.

However, when we reach these very high energies we need to rethink what temperature really means. As we continually increase the temperature of a gas by giving it more energy, the electrons may eventually be stripped from the atoms turning the gas into a plasma. Plasma is essentially a gas of electrons mixed with a gas of atomic nuclei, and eventually the amount of energy each particle has may mean it escapes whatever means we have to contain it, usually an electric or magnetic field. So how do we measure its temperature then?

To measure temperature, we usually bring one system into contact with another and wait until they reach thermal equilibrium – you put a mercury thermometer under your tongue until the mercury stops rising, and that is your temperature. Temperature is really a measure of heat transfer between two objects.

We can measure the temperature of a gas, or even a plasma, in terms of the energy it transfers to another object before they come into thermal equilibrium, which is fine until we give so much energy to the plasma that the individual particles escape. If we measure the energy released in a collision between one of these particles and another, are we measuring the temperature of that particle or simply the energy change in the collision? This is why we tend to forget the idea of temperature at this kind of level and just use it in relation to gases, liquid or solids.

Can we ever reach
absolute zero?

Absolute zero is the temperature at which all molecules and atoms cease to moved. It is minus 273.15°C. We cannot ever reach absolute zero. At absolute zero all the molecules or atoms of a substance would not be moving at all. The third law of thermodynamics prevents us from getting there.

The scientist and author C.P. Snow had an excellent way of expressing the three laws:

1. You cannot win (that is, you cannot get something for nothing, because matter and energy are conserved).

2. You cannot break even (you cannot return to the same energy state, because there is always an increase in disorder; entropy always increases).

3. You cannot get out of the game (because absolute zero is unattainable).

So we can never get to absolute zero, but in 1995 a team of American physicists conducted an experiment and claimed to have got within a billionth of a degree of it.

What causes **gravity?**

This is perhaps the hardest question we've ever been asked, and the answer is probably the hardest to understand.

Our idea of gravity today is based on Einstein's General Theory of Relativity. He said that space-time is warped and stretched by matter. But what is space-time?

Space-time is neither space nor time, but a mix of the two. In physics, you describe an event by saying where and when it happened. But, *when* something happened depends on *where* you are. In everyday life you don't really notice it, but if you were on a spacecraft which travelled at speeds close to the speed of light, then you would notice that your clock would not match that of a friend who stayed stuck on Earth. Their clock would appear to be running more slowly. At the moment it's all theory, but suppose we ever did get to fly around the galaxy at high speed, this would become very important. You'll say to your children: 'Hey kids! We've reached

the star Sirius so put your digital watches forward 8 years.' To you and your children the journey may have felt like it lasted a few weeks, but to people on the Earth and Sirius it would have felt that your journey had taken nearly eight years.

So, when you talk about space-time you have to be careful that you don't separate space and time like you do in everyday life.

Now, back to gravity. Everything which is free-falling is just following a straight line in space-time. Straight lines in space are not necessarily straight lines in space-time. Imagine drawing a straight line on a flat, two-dimensional map. If you tried following that path in the true, three-dimensional world in which we live, it probably wouldn't be a straight line at all – it would go up hills and down valleys. The straight line in 2D could be a squiggle in 3D. The same happens in space-time. When satellites orbit the Earth their path in space is curved, but their path in space-time is a straight line. That's why objects appear to be attracted to the Earth; they curve towards the Earth in space but they're just trying to follow straight lines in space-time.

But how does mass distort space-time in this way? Well, space and time don't exist without matter, and we'll prove it. How do you measure time? The answer is by the rate of change of something such as the swing of a pendulum, or the pulse of quartz crystal in your digital watch. All these things are made of matter. *Time is a measure of change, but all change involves matter. You can't have time without matter.*

So how do you measure space? With a ruler. But rulers are made of matter and so tied up with space, time *and space-time, too.* Phew!

Let's see if I can express this in one sentence: objects acting under the influence of gravity travel in straight lines in space-time, but matter bends space-time so objects appear to fall towards other matter.

Can the value of **pi** ever **change?**

No. In mathematics, the symbol π denotes the ratio of the circumference of a circle to its diameter. The ratio is approximately 3.14159265, but the string of numbers goes way beyond that, possibly to infinity. Pi is called an irrational number because it cannot be expressed as a simple fraction, or as a decimal with a finite number of decimal places. It is also a transcendental number which means it has no continuously recurring digits. Electronic computers in the late twentieth century have carried pi to more than 200 billion decimal places and proved this to be the case.

The only way pi can change is by the accuracy to which it is calculated.

In very ancient times, three was used as the approximate value of pi, and not until Archimedes (3rd century BC) does there seem to have been a scientific effort to compute it – he

reached a figure equivalent to about 3.14. A figure equivalent to 3.1416 dates from before AD200. By the early sixth century Chinese and Indian mathematicians had independently confirmed or improved the number of decimal places. By the end of the seventeenth century in Europe, new methods of mathematical analysis provided various ways of calculating pi.

Early in the twentieth century the Indian mathematical genius Srinivasa Ramanujan developed ways of calculating pi that were so efficient that they have been incorporated into computer algorithms, giving pi to millions of digits.

And finally,
one of the most **important** questions
Science Line was ever asked:
Can two people hold **totally
opposing** views on **science**,
and both be **right?**

Yes. Quite easily. Science is all about collecting data and interpreting it. Collecting the data might be objective, but interpreting it certainly isn't. And that's the whole thing with science: you collect data until you've got enough from lots of different sources that convinces everyone to agree with your interpretation of it, but there are many different interpretations along the way.

We will probably never know everything there is to know, so even the answers in this book which might appear to be the final word on the subject at the beginning of the twenty-first century, are merely assumptions we make as we travel the long road towards ultimate knowledge.

Index

242